D1549468

WITHDRAWN FROM STOCK

BrightRED Revision

Advanced

Higher MATHS

Dr Michael Green,
Linda Moon, Philip Moon

BrightRED
PUBLISHING

First published in 2011 by:
Bright Red Publishing Ltd
6 Stafford Street
Edinburgh
EH3 7AU

Copyright © Bright Red Publishing Ltd 2011

Cover image © Caleb Rutherford

All rights reserved. No part of this publication may be reproduced, stored in a retrieval system, or transmitted in any form or by any means, electronic, mechanical, photocopying, recording or otherwise, without prior permission in writing from the publisher.

The rights of Michael Green, Linda Moon and Philip Moon to be identified as the authors of this work have been asserted by them in accordance with sections 77 and 78 of the Copyright, Designs and Patents Act 1988

A CIP record for this book is available from the British Library

ISBN 978-1-906736-29-3

With thanks to:
The Partnership Publishing Solutions (layout), ARTLIFE (illustrations), Tony Wayte (copy-edit) and Ivor Normand (proof-read)

Cover design by Caleb Rutherford – e i d e t i c

Acknowledgements
Every effort has been made to seek all copyright holders. If any have been overlooked, then Bright Red Publishing will be delighted to make the necessary arrangements.

All internet links in the text were correct at the time of going to press.

Bright Red Publishing would like to thank the Scottish Qualifications Authority for use of Past Exam Questions. Answers do not emanate from SQA.

Printed and bound in Scotland by Bell & Bain Limited, Glasgow.

Mixed Sources
Product group from well-managed
forests and other controlled sources
www.fsc.org Cert no. TT-COC-002769
© 1996 Forest Stewardship Council
FSC

LEICESTER LIBRARIES	
Askews & Holts	20-Apr-2011
	£14.99

CONTENTS

INTRODUCTION

COURSE STRUCTURE

The Advanced Higher Mathematics course is divided into three units:

Unit 1:

- Algebraic Skills
- Rules of Differentiation
- Integration
- Properties of Functions
- Systems of Linear Equations

Unit 2:

- Further Differentiation
- Further Integration
- Complex Numbers
- Sequences and Series
- Number Theory and Proof

Unit 3:

- Vectors in Three Dimensions
- Matrix Algebra
- Further Sequences and Series
- Further Differential Equations
- Further Number Theory and Proof

ASSESSMENT

The Advanced Higher Mathematics course is assessed in two ways:

- Each of the five outcomes for the three units is assessed within your school using a National Assessment Bank test. NABs are set by the Scottish Qualifications Authority and consist of structured, short-answer questions at grade C. The percentage required for a pass varies for each outcome between 60% (3 marks out of 5) and 75% (9 marks out of 12).

- You will also take an externally assessed written examination consisting of a paper lasting three hours. The examination has an allocation of 100 marks. It is made up of shorter questions, usually twelve, that are generally worth between 3 and 7 marks. The final four questions will be longer and worth 40 marks in total.

EXAM HINTS

You do not need to answer the questions in order. Find a question that you can answer easily, so that you settle your nerves. However, remember that the examiners will have attempted to put the shorter questions in order of difficulty, followed by the longer ones, also in order of difficulty. Thus question 12 might well be harder than question 16.

Remember: 100 marks in three hours, that is 1·8 minutes for every mark, or 18 minutes for a long question.

The course is graded A bands 1 and 2, B bands 3 and 4, C bands 5 and 6, or D band based on how well you do in the external examination. To gain a course award, you must also pass the three NABs (one for each unit).

THE STRUCTURE AND AIM OF THIS BOOK

There is no short-cut to passing any course at this Advanced Higher level. To obtain a good pass requires consistent, regular revision over the duration of the course. The aim of this revision book is to bring together, in one volume, concise coverage of the course material. The book should be used in conjunction with your course notes and the knowledge you gain from attending classes.

In addition to an index at the end of the book, there is a summary contents list which details, in order of the syllabus given by SQA, each thread of the course – unit by unit, outcome by outcome. There is a box alongside, which you might use to monitor and chart your progress, perhaps as follows:

☐ an empty box: I have not learned this

• a dot: we have covered this in class

– a dash: I understand this topic, but more practice is required

+ a cross: I have revised this area and I am confident with it.

THE STRUCTURE AND AIM OF THIS BOOK   contd

Regularly throughout this book are:

- **Don't Forget** boxes that flag up vital pieces of information that you need to remember, and important things that you must be able to do, plus some helpful hints.

- **Now Try This** sections that contain practice questions to test your understanding. The solutions to these tasks are on the Bright Red Publishing website (www.brightredpublishing.co.uk).

- **Websites**: internet links to use to extend your knowledge of the subject. In addition, www.hsn.uk.net has summary notes, and the SQA website (www.sqa.org.uk) has information on the exam which may be useful.

REVISION TIPS

General advice:

- Don't leave your revision until the last minute. When you are still learning new topics, revise the ones you have already covered.

- Study for periods of between 30 and 45 minutes, unless you are doing a complete paper.

- Take short breaks, away from your study area, to keep your level of concentration high.

- During your study leave, build treats and relaxation time into your revision timetable. This will help you to focus and help you stick to your plan.

In the run up to the exams, Eat Well, Exercise Well and Sleep Well.

Maths-specific:

- The best way to revise mathematics is by doing it. There is a time for learning the necessary formulae and rules, but there is no substitute for practice.

- Once you have learned a topic or skill, try questions. Start off with straightforward questions, then NAB level, and progress to examination style. You want to test your knowledge of a topic by trying that discrete area of the syllabus, but you should progress to trying a mixture of past-paper questions. It is important that you start to recognise what to use and when, which skill to apply and where.

- Use the space between **Don't Forgets** in the scholar's margin to add your own revision reminders.

- Mathematics is a subject to be practised often. Try to get into the habit of regularly doing mathematics. If you complete one extra question every night in addition to your normal homework and study time at school, you will reap the rewards. You will be able to ask for help the next day when the problem is fresh in your mind, so that you can tackle another question the next evening – and so you will quickly build up your knowledge and confidence.

- Mathematics also demands perseverance – and there will come a time when you need to tackle a number of questions or a whole examination paper in one sitting. Time management is essential.

- Mathematics is different from other subjects in so many ways and the good thing about revising it is that you can be active.

The Best Way to Revise Maths is to Actively Do It.

FACTORIALS AND BINOMIAL COEFFICIENTS

When n is a positive integer, the **factorial** of n is denoted by $n!$, where:

$$n! = n \times (n-1) \times (n-2) \times \ldots \times 2 \times 1$$

Your calculator should have an $n!$ button, but because the factorial increases dramatically it probably won't be able to calculate factorials above $n = 69$.

DON'T FORGET

$n! = n \times (n-1)!$
0! is defined to be 1.

Example 1.1

$4! = 4 \times 3 \times 2 \times 1 = 24$

THE BINOMIAL COEFFICIENT $\binom{n}{r}$

When n is a positive integer, and r is an integer with $0 \leqslant r \leqslant n$, $\binom{n}{r} = \dfrac{n!}{r!(n-r)!}$ where $\binom{n}{r}$ is called the **binomial coefficient**.

DON'T FORGET

$\binom{n}{r} = \binom{n}{n-r}$

Example 1.2

$\binom{5}{2} = \dfrac{5!}{2!3!} = \dfrac{5 \times 4 \times 3 \times 2 \times 1}{2 \times 1 \times 3 \times 2 \times 1} = 10$

An alternative notation for the binomial coefficient $\binom{n}{r}$ is ^nC_r, or $_nC_r$. Check your calculator to see which notation is used.

DON'T FORGET

$\binom{n}{0} = \binom{n}{n} = 1$ for all n

If your calculator doesn't have factorial or binomial coefficient buttons (or if it packs up during the examination), there will always be factors which cancel to give a whole number.

Example 1.3

$\binom{25}{3} = \dfrac{25!}{3!\,22!} = \dfrac{25 \times 24 \times 23 \times 22 \times \ldots}{3 \times 2 \times 1 \times 22 \times \ldots} = \dfrac{25 \times 24 \times 23 \times 22!}{6 \times 22!} = 4 \times 25 \times 23 = 2300$

DON'T FORGET

$\binom{n}{2} = \dfrac{1}{2}n(n-1)$

Example 1.4

Solve the equation $\binom{n}{3} = 2\binom{n}{2}$

Method 1

$$\dfrac{n!}{3!(n-3)!} = \dfrac{2 \times n!}{2!\,(n-2)!}$$

so $(n-2)! = 3!(n-3)!$

$(n-2)(n-3)! = 6(n-3)!$

$n - 2 = 6$

$n = 8$

contd

THE BINOMIAL COEFFICIENT   contd

Method 2

$$\binom{n}{3} = \frac{1}{6}n(n-1)(n-2) \text{ and } \binom{n}{2} = \frac{n(n-1)}{2}$$

so, $\frac{1}{6}n(n-1)(n-2) = n(n-1)$.

Dividing by $n(n-1)$, which is non-zero because $n \geqslant 3$ for $\binom{n}{3}$ to be defined, gives:

$$\frac{n-2}{6} = 1 \implies n = 8.$$

⚙ NOW TRY THIS

(1) Show that $\binom{n+1}{3} - \binom{n}{3} = \binom{n}{2}$

where the integer n is greater than or equal to 3.   4

(2) Solve the equation $2\binom{n}{4} = \binom{n}{2}$.   4

(3) Prove that

$$\binom{n}{n-3} + \binom{n+1}{3} = \frac{1}{6}n(n-1)(2n-1).$$

5

THE BINOMIAL THEOREM

DON'T FORGET

The powers of a and b in the expansion always add up to n.

DON'T FORGET

Using Σ notation for summation,

$(a+b)^n = \sum_{r=0}^{n} \binom{n}{r} a^{n-r} b^r$

DON'T FORGET

Remember to simplify the coefficients in a binomial expansion.

DON'T FORGET

The general term is

$\binom{n}{r} a^{n-r} b^r$

DON'T FORGET

'The term independent of x' means that the term you are trying to find will not contain x.

The binomial theorem states that:

$$(a+b)^n = a^n + \binom{n}{1}a^{n-1}b + \binom{n}{2}a^{n-2}b^2 + \dots + \binom{n}{n-1}ab^{n-1} + b^n$$

for any positive integer n. The binomial theorem is used to describe the expansion of powers of a binomial, using a sum of terms. Coefficients in the expansion are binomial coefficients.

Example 1.5

Obtain the binomial expansion of $(x + 2)^5$.

$$(x+2)^5 = x^5 + \binom{5}{1}x^4 2 + \binom{5}{2}x^3 2^2 + \binom{5}{3}x^2 2^3 + \binom{5}{4}x 2^4 + 2^5$$

$$(x+2)^5 = x^5 + 10x^4 + 40x^3 + 80x^2 + 80x + 32.$$

Example 1.6

Obtain the binomial expansion of $(2a - 3)^4$.

$$(2a-3)^4 = {}^4C_0(2a)^4(-3)^0 + {}^4C_1(2a)^3(-3)^1 + {}^4C_2(2a)^2(-3)^2 + {}^4C_3(2a)^1(-3)^3 + {}^4C_4(2a)^0(-3)^4$$

$$= (2a)^4 + 4(2a)^3(-3) + 6(2a)^2(-3)^2 + 4(2a)(-3)^3 + (-3)^4$$

$$= 16a^4 - 96a^3 + 216a^2 - 216a + 81.$$

Note that, when you are asked to write out the full expansion in assessments, the expansion power will be no higher than 5.

Example 1.7

Obtain the term independent of x in the expansion of $\left(x + \dfrac{3}{x} \right)^4$.

The general term is:

$$\binom{4}{r}x^{4-r}\left(\frac{3}{x}\right)^r = \binom{4}{r}3^r x^{4-2r}.$$

We want the power of x to be 0, so:

$$4 - 2r = 0,\ r = 2.$$

Thus the constant term is:

$$\binom{4}{2}3^2 = 6 \times 9 = 54.$$

Example 1.8

Obtain the coefficient of x^2 in the expansion of $\left(3x^2 - \dfrac{1}{x^3} \right)^{11}$.

The general term is:

$$\binom{11}{r}(3x^2)^{11-r}\left(-\frac{1}{x^3}\right)^r = \binom{11}{r}3^{11-r}(-1)^r x^{22-5r}.$$

We need $22 - 5r = 2 \implies r = 4$.

Hence the coefficient of x^2 is:

$$\binom{11}{4}3^{11-4}(-1)^4 = \frac{11 \times 10 \times 9 \times 8 \times 3^7}{1 \times 2 \times 3 \times 4} = 721710.$$

PASCAL'S TRIANGLE

 http://en.wikipedia.org/wiki/Pascal%27s_triangle
http://planetmath.org/encyclopedia/PascalsTriangle.html

Pascal's triangle is a triangular array of binomial coefficients.

Example 1.9

The first few rows of Pascal's triangle are shown here:

$$
\begin{array}{ccccccccc}
 & & & & 1 & & & & \\
 & & & 1 & & 1 & & & \\
 & & 1 & & 2 & & 1 & & \\
 & 1 & & 3 & & 3 & & 1 & \\
1 & & 4 & & 6 & & 4 & & 1
\end{array}
$$

Pascal's triangle depends on the result $\binom{n+1}{r} = \binom{n}{r} + \binom{n}{r-1}$. You should know how to prove this result.

Note that $\binom{4}{0} = 1$ $\binom{4}{1} = 4$ $\binom{4}{2} = 6$ $\binom{4}{3} = 4$ $\binom{4}{4} = 1$.

DON'T FORGET

$$\binom{n+1}{r} = \binom{n}{r} + \binom{n}{r-1}$$

NOW TRY THIS

(1) Expand $(a^3 + 2)^4$ by the binomial theorem.   3

(2) Write down and simplify the general term in the expansion of $\left(x^2 + \dfrac{1}{x}\right)^{10}$.   3

Hence, or otherwise, obtain the term in x^{14}.   2

(3) Find constants A, B, C such that

$$\left(z - \frac{1}{z}\right)^5 = A\left(z^5 - \frac{1}{z^5}\right) + B\left(z^3 - \frac{1}{z^3}\right) + C\left(z - \frac{1}{z}\right).$$ 3

(4) Write down the general term in the binomial expansion of $\left(3x - \dfrac{2}{x^2}\right)^6$.

Hence obtain the term independent of x.   5

(5) By writing $(1 + x)^{n+1} = (1 + x)^n (1 + x)$ and considering the coefficient of $x^r (1 \leqslant r \leqslant n)$ on both sides, prove that

$$\binom{n+1}{r} = \binom{n}{r} + \binom{n}{r-1}.$$ 4

(6) Prove that

$$\binom{n}{r} + 2\binom{n}{r+1} + \binom{n}{r+2} = \binom{n+2}{r+2}.$$ 4

(7) Prove that

$$\binom{n}{1} + \binom{n}{2} + \dots + \binom{n}{n-1} = 2^n - 2.$$ 4

PARTIAL FRACTIONS

Certain types of rational functions $\dfrac{p(x)}{q(x)}$, where p and q are polynomials in x, can be decomposed into partial fractions. This can be useful for integrating or differentiating this type of function. For examination purposes, q can be:

- quadratics or cubics which can easily be factorised into linear factors, or
- cubics which can be factorised into a product of a linear factor and an irreducible quadratic factor.

Example 1.10

Express $\dfrac{x-5}{x^2-x-2}$ in partial fractions.

$$\frac{x-5}{x^2-x-2} = \frac{x-5}{(x+1)(x-2)} = \frac{A}{x+1} + \frac{B}{x-2}.$$

Multiplying through by $(x+1)(x-2)$ gives $x - 5 = A(x-2) + B(x+1)$.
$x = 2$ gives $B = -1$, $\quad x = -1$ gives $A = 2$, so:

$$\frac{x-5}{x^2-x-2} = \frac{2}{x+1} - \frac{1}{x-2}.$$

To obtain the partial fraction decomposition of $\dfrac{p(x)}{q(x)}$ when the degree of p is $\geqslant$ the degree of q, use long division to obtain a remainder of the form $\dfrac{r(x)}{q(x)}$,

where the degree of r is $<$ the degree of q.

Partial fractions can now be performed on $\dfrac{r(x)}{q(x)}$.

DON'T FORGET

Long division is required

for $\dfrac{p(x)}{q(x)}$ when

deg $p(x) \geqslant$ deg $q(x)$.

Example 1.11

$\dfrac{2x^2-x-9}{x^2-x-2} = 2 + \dfrac{x-5}{x^2-x-2}$ using long division;

now proceeding as in Example 1.10, we get:

$$\frac{2x^2-x-9}{x^2-x-2} = 2 + \frac{2}{x+1} - \frac{1}{x-2}.$$

Long division

$$\begin{array}{r} 2 \\ x^2-x-2\overline{\smash{)}2x^2-x-9} \\ \underline{2x^2-2x-4} \\ x-5 \end{array}$$

First divide $2x^2$ by x^2, then multiply the **whole** of $x^2 - x - 2$ (the divisor) by 2.

Now subtract the result from $2x^2 - x - 9$ to get $x - 5$.

Because the highest power of x is now less than 2, the process stops, giving the result shown.

Example 1.12   Repeated factor

$$\frac{3x^2 - 11x + 4}{(x+1)(x-2)^2} = \frac{A}{x+1} + \frac{B}{x-2} + \frac{C}{(x-2)^2}$$

$$3x^2 - 11x + 4 = A(x-2)^2 + B(x+1)(x-2) + C(x+1)$$

$x = -1$ gives $A = 2$, $x = 2$ gives $C = -2$.

Comparing the coefficient of x^2 on both sides gives
$3 = A + B$, from which $B = 1$.

Finally, $\dfrac{3x^2 - 11x + 4}{(x+1)(x-2)^2} = \dfrac{2}{x+1} + \dfrac{1}{x-2} - \dfrac{2}{(x-2)^2}$.

> **DON'T FORGET**
>
> If $q(x)$ contains a factor $(x - a)^2$, you need
> $$\frac{A}{x-a} + \frac{B}{(x-a)^2}$$

If $q(x)$ contains an irreducible factor such as $x^2 + a^2$
you need $\dfrac{Ax+B}{x^2+a^2}$, or

$x^2 + x + 1$ you need $\dfrac{Ax+B}{x^2+x+1}$.

Example 1.13   Irreducible factor

$$\frac{2x^2 + x + 27}{x^3 + 9x} = \frac{A}{x} + \frac{Bx+C}{x^2+9}, \text{ so:}$$

$$2x^2 + x + 27 = A(x^2+9) + x(Bx+C).$$

Setting $x = 0$ gives $A = 3$.

Comparing the coefficient of x^2 on both sides gives $2 = A + B$, so $B = -1$.

Comparing the coefficient of x on both sides gives $C = 1$, so:

$$\frac{2x^2 + x + 27}{x^3 + 9x} = \frac{3}{x} - \frac{x-1}{x^2+9}.$$

> **DON'T FORGET**
>
> Use $Ax + B$ for irreducible quadratics.

NOW TRY THIS

(1) Express $\dfrac{x-2}{3x^2 + 10x + 3}$ in partial fractions.   2

(2) Express $\dfrac{1}{x^3 + 4x}$ in partial fractions.   4

(3) Express $\dfrac{x^3 + x^2 + 2}{(x+1)^2}$ in partial fractions.   5

STANDARD DIFFERENTIALS

You should know all of the differentials in the following table or be able to derive them. For example: $\frac{d}{dx}(\tan x) = \frac{d}{dx}\left(\frac{\sin x}{\cos x}\right)$ and then use the quotient rule (or the product rule) as below.

DERIVATIVES TO LEARN

> **DON'T FORGET**
>
> You should be able to recognise and use different notations:
> - functional notation: $f(x)$, $f'(x)$, $f''(x)$
> - Leibniz notation: $\frac{dy}{dx}$, $\frac{d^2y}{dx^2}$

> **DON'T FORGET**
>
> e^{3x} can be written as $\exp(3x)$

> **DON'T FORGET**
>
> $\tan^{-1}x$ is **not** $\dfrac{1}{\tan x}$

$f(x)$	$f'(x)$	
e^x	e^x	(Unit 1)
$\ln x$ $(or \log_e x)$	$\dfrac{1}{x}$	(Unit 1)
$\tan x$	$\sec^2 x$	(Unit 1)
$\sec x$	$\sec x \tan x$	(Unit 1)
$\operatorname{cosec} x$	$-\dfrac{\cos x}{\sin^2 x} = -\cot x \operatorname{cosec} x$	(Unit 1)
$\cot x$	$-\operatorname{cosec}^2 x$	(Unit 1)
$\sin^{-1} x$, $\sin^{-1} ax$	$\dfrac{1}{\sqrt{1-x^2}}$, $\dfrac{a}{\sqrt{1-(ax)^2}}$	(Unit 2)
$\cos^{-1} x$, $\cos^{-1} ax$	$\dfrac{-1}{\sqrt{1-x^2}}$, $\dfrac{-a}{\sqrt{1-(ax)^2}}$	(Unit 2)
$\tan^{-1} x$, $\tan^{-1} ax$	$\dfrac{1}{1+x^2}$, $\dfrac{a}{1+(ax)^2}$	(Unit 2)

As with $\sin x$, $\cos x$ from Higher, these new trig derivative results require x to be expressed in radians.

RULES FOR DIFFERENTIATION

> **DON'T FORGET**
>
> $f'g + fg'$

Product rule

The product rule can be expressed as:

$$\left(f(x)g(x)\right)' = f'(x)g(x) + f(x)g'(x)$$

and is used to find the derivatives of products of functions.

Example 2.1

$$\frac{d}{dx}(4x \sin x) = 4\sin x + 4x \cos x$$

> **DON'T FORGET**
>
> $\dfrac{f'g - fg'}{g^2}$

Quotient rule

The quotient rule can be expressed as:

$$\left(\frac{f(x)}{g(x)}\right)' = \frac{f'(x)g(x) - f(x)g'(x)}{\left(g(x)\right)^2}$$

and is used to find the derivatives of quotients of functions.

> **DON'T FORGET**
>
> Remember to square the denominator $g(x)$

Example 2.2

$$\frac{d}{dx}\left(\frac{\sin x}{\ln x}\right) = \frac{\cos x \ln x - \sin x \cdot \frac{1}{x}}{\left(\ln x\right)^2} = \frac{x\cos x \ln x - \sin x}{x\left(\ln x\right)^2}$$

contd

RULES FOR DIFFERENTIATION   contd

Chain rule

The chain rule can be expressed as: $(f(g(x)))' = f'(g(x)) \cdot g'(x)$

and is used for differentiating a composition of functions.

Example 2.3

$$\frac{d}{dx}(e^{\tan x}) = e^{\tan x} \cdot \sec^2 x$$

Example 2.4

$$\frac{d}{dx}\left(\tan^{-1}(1-2x)\right) = \frac{1}{1+(1-2x)^2} \times (-2)$$

$$= \frac{-2}{2-4x+4x^2} = -\frac{1}{1-2x+2x^2}$$

DON'T FORGET

Tidy up solutions.

DON'T FORGET

Remember $\dfrac{dy}{dx} = \dfrac{1}{\dfrac{dx}{dy}}$

DON'T FORGET

Remember

$$f'(x) = \lim_{h \to 0} \frac{f(x+h)-f(x)}{h}$$

HIGHER DERIVATIVES

If y is a function of x, so is $\dfrac{dy}{dx}$, and this can also be differentiated with respect to x. We denote $\dfrac{d}{dx}\left(\dfrac{dy}{dx}\right)$ by $\dfrac{d^2y}{dx^2}$, and call this the second derivative of y with respect to x. In the same way, we can form the third derivative, $\dfrac{d^3y}{dx^3}$ and so on. If we are using the notation $f'(x)$, then the second derivative is denoted by $f''(x)$ and so on.

Example 2.5

$$\frac{d^{n+1}y}{dx^{n+1}} = \frac{d}{dx}\left(\frac{d^n y}{dx^n}\right)$$

$$y = (3x+1)^4$$

$$\frac{dy}{dx} = 12(3x+1)^3$$

$$\frac{d^2y}{dx^2} = 108(3x+1)^2.$$

Example 2.6

If $\dfrac{d^2y}{dx^2} = \sin^4 2x$,

then $\dfrac{d^3y}{dx^3} = 8\cos 2x \sin^3 2x$.

DON'T FORGET

Give exact values in your answers rather than decimal approximations.

Example 2.7

A car travels along a straight road starting from rest at time $t = 0$ at a point A. After t seconds, it has travelled a distance $\dfrac{t^3}{3} + t^2$ metres. At what time will the car break a speed limit of $25\,\text{ms}^{-1}$?

Let $x = \dfrac{t^3}{3} + t^2$ be the distance travelled, so that the speed $\dfrac{dx}{dt} = t^2 + 2t$. We need the positive value of t for which $t^2 + 2t = 25$. This gives $t = \sqrt{26} - 1 = 4 \cdot 1\,\text{s}$ to 1 decimal place.

The car will break the speed limit after $4 \cdot 1\,\text{s}$. (Maxima/minima questions appear in Chapter 4.)

DON'T FORGET

Any of these differentiation skills could be tested in contextualised questions such as optimisation or displacement/velocity/acceleration.

NOW TRY THIS

(1) Define $f(x) = \sin^2 x\, e^{-\tan x}$. Obtain $f'(x)$ and evaluate $f'(\frac{\pi}{4})$.   **3, 1**

(2) Given $f(x) = x^2 \tan 3x\ (0 < x < \frac{\pi}{6})$, obtain $f'(x)$.   **3**

(3) Given $y = \dfrac{e^x}{1+2x}\ (x \neq -\frac{1}{2})$, obtain the value of x for which $\dfrac{dy}{dx} = 0$.   **3**

(4) The amount x micrograms of an impurity removed per kg of a substance by a chemical process depends on the temperature $T\,^\circ$C as follows:

$x = T^3 - 90T^2 + 2400T, \quad 10 \leqslant T \leqslant 60.$

At what temperature in the given range should the process be carried out to remove as much impurity per kg as possible?   **4**

FURTHER DIFFERENTIATION 1

DERIVATIVES OF INVERSE FUNCTIONS

Let $y = f(x)$ be a function of x such that each value of y is given by a unique value of x. Then x can be regarded as a function of y:

$x = f^{-1}(y)$. This is the **inverse function** of f. (Inverse functions are discussed in greater detail in Chapter 4.)

To obtain $\frac{d}{dx}[f^{-1}(x)]$ for a given f, proceed as follows:

Let $y = f^{-1}(x)$ so that $x = f(y)$ and $\frac{dx}{dy} = f'(y) \implies \frac{dy}{dx} = \frac{1}{f'(y)}$.

Now express this in terms of x, simplifying where possible.

Example 2.8

Obtain the derivative of $\sec^{-1} x$, where $x \geqslant 1$.

$y = \sec^{-1} x \implies x = \sec y$, so $\frac{dx}{dy} = \sec y \tan y$. Using $\tan y = \sqrt{\sec^2 y - 1}$ gives:

$\frac{dx}{dy} = x\sqrt{x^2 - 1}$ so that $\frac{d}{dx}(\sec^{-1} x) = \frac{1}{x\sqrt{x^2 - 1}}$.

DON'T FORGET

$f^{-1}(x)$ is not the same as $\frac{1}{f(x)}$

IMPLICIT DIFFERENTIATION

An equation of the form $f(x,y) = 0$ defines a curve in the xy-plane. This may define y as a function of x, or more than one function of x.

For example, the equation $x^2 + x^2 - 4 = 0$ defines a circle with centre the origin and radius 2.

It defines two functions, $y = +\sqrt{4 - x^2}$ and $y = -\sqrt{4 - x^2}$ for $-2 \leqslant x \leqslant 2$.

Given $f(x,y) = 0$, to obtain $\frac{dy}{dx}$ we use implicit differentiation.

DON'T FORGET

If y is a function of x, then $\frac{d}{dx}[g(y)] = g'(y)\frac{dy}{dx}$, i.e. the chain rule.

Example 2.9

The equation $x^4 + x^2y^2 + y^4 = 1$ defines y implicitly as a function of x. Obtain $\frac{dy}{dx}$.

Using the product rule and chain rule on the second term, and the chain rule on the third term,

$4x^3 + 2xy^2 + 2x^2y\frac{dy}{dx} + 4y^3\frac{dy}{dx} = 0$.

Gathering terms involving $\frac{dy}{dx}$ gives: $\frac{dy}{dx}(2x^2y + 4y^3) = -4x^3 - 2xy^2$.

Cancelling a factor of 2 gives: $\frac{dy}{dx} = -\frac{2x^3 + xy^2}{2y^3 + x^2y}$.   Eqn 1

Finding the 2nd derivative gives:

$\frac{d^2y}{dx^2} = -\frac{(2y^3 + x^2y)(6x^2 + y^2 + 2xy\frac{dy}{dx}) - (2x^3 + xy^2)(6y^2\frac{dy}{dx} + 2xy + x^2\frac{dy}{dx})}{(2y^3 + x^2y)^2}$.

Unless you are specifically asked for an expression for $\frac{d^2y}{dx^2}$ in terms of x and y, do not replace $\frac{dy}{dx}$ in this expression by its result from Eqn 1 above. The most likely information you will need from the 2nd derivative is its numerical value for a given (x,y), so simply substitute for x and y in each of the separate derivative expressions.

EQUATIONS OF TANGENTS TO CURVES DEFINED IMPLICITLY

Example 2.10

Obtain an equation for the tangent at the point (1, 2) on the curve defined by:

$$xy^3 - 2x^2y^2 + x^4 - 1 = 0.$$

$\frac{d}{dx}(xy^3) - \frac{d}{dx}(2x^2y^2) + 4x^3 = 0$, so:

$$y^3 + 3xy^2\frac{dy}{dx} - 4xy^2 - 4x^2y\frac{dy}{dx} + 4x^3 = 0.$$

Let the gradient of the tangent at the point (1, 2) be m.

Then $8 + 12m - 16 - 8m + 4 = 0$, giving $m = 1$, and an equation for the tangent at (1, 2) is $y = x + 1$.

NOW TRY THIS

(1) Differentiate with respect to x

    (a) $f(x) = (2 + x)\tan^{-1}\sqrt{x - 1}, \ x > 1$ 4

    (b) $g(x) = e^{\cot 2x}, \ 0 < x < \dfrac{\pi}{2}$ 2

(2) The equation $y^2 + 3xy = x^2 - 3$ defines a curve through the point $P(-1, 1)$.
Obtain the gradient of the tangent to the curve at P.   4

(3) Given $xy - x = 4$, use implicit differentiation to obtain $\dfrac{dy}{dx}$ in terms of x and y.   2

    Hence obtain $\dfrac{d^2y}{dx^2}$ in terms of x and y.   3

FURTHER DIFFERENTIATION 2

LOGARITHMIC DIFFERENTIATION

DON'T FORGET

With functions of the form $f(x)^{g(x)}$, as in Example 2.11, logarithmic differentiation **must** be used.

DON'T FORGET

If the derivative of a function appears to involve several combinations of the product, quotient or chain rules, then logarithmic differentiation may be best.

Logarithmic differentiation refers to the process of first taking logs (to base e), then differentiating the result of this.

Example 2.11

Obtain $\dfrac{d}{dx}(x^{\cos x})$, where $x > 0$.

Let $y = x^{\cos x} \implies \ln y = \cos x \ln x$, so:

$$\frac{1}{y}\frac{dy}{dx} = \frac{\cos x}{x} - \sin x \ln x, \text{ which gives:}$$

$$\frac{dy}{dx} = x^{\cos x - 1} \cos x - x^{\cos x} \sin x \ln x.$$

Example 2.12

Obtain $\dfrac{dy}{dx}$ when $y = \sqrt{\dfrac{1+x^2}{1-x^2}}$, where $-1 < x < 1$.

$\ln y = \dfrac{1}{2}\ln(1+x^2) - \dfrac{1}{2}\ln(1-x^2)$, so:

$$\frac{1}{y}\frac{dy}{dx} = \frac{x}{1+x^2} + \frac{x}{1-x^2} = \frac{2x}{(1+x^2)(1-x^2)}$$

$$\frac{dy}{dx} = \frac{2xy}{(1+x^2)(1-x^2)} = \frac{2x\sqrt{1+x^2}}{(1+x^2)(1-x^2)\sqrt{1-x^2}} = \frac{2x}{(1+x^2)^{1/2}(1-x^2)^{3/2}}.$$

PARAMETRIC DIFFERENTIATION

Equations of the form $x = f(t)$, $y = g(t)$, where t takes values in some interval, describe a curve in the xy-plane. This curve may define a single function y of x, or more than one function. These equations are called **parametric equations**.

DON'T FORGET

If x and y are functions of t,

$\dfrac{dy}{dx} = \dfrac{dy}{dt} \bigg/ \dfrac{dx}{dt}$, and

$\dfrac{d^2y}{dx^2} = \dfrac{d}{dt}\left(\dfrac{dy}{dx}\right) \bigg/ \dfrac{dx}{dt}$

For example, $x = t + 1$, $y = t^2$, where t is any real number. Here it is easy to eliminate t to give $y = (x - 1)^2$. However, in the examination, you shouldn't eliminate unless asked to.

We often need to obtain $\dfrac{dy}{dx}$ (and sometimes $\dfrac{d^2y}{dx^2}$) when x and y are given parametrically.

Example 2.13

Obtain $\dfrac{dy}{dx}$ and $\dfrac{d^2y}{dx^2}$ given that $x = t^3 + t$, $y = t^2 + 1$.

$$\frac{dx}{dt} = 3t^2 + 1, \quad \frac{dy}{dt} = 2t$$

so, $\dfrac{dy}{dx} = \dfrac{2t}{3t^2 + 1}$

$$\frac{d^2y}{dx^2} = \frac{d}{dx}\left(\frac{dy}{dx}\right) = \frac{d}{dt}\left(\frac{2t}{3t^2+1}\right)\frac{dt}{dx}$$

$$= \frac{2 - 6t^2}{(3t^2+1)^2} \times \frac{1}{3t^2+1} = \frac{2-6t^2}{(3t^2+1)^3}.$$

RELATED RATES OF CHANGE

You might be given a problem involving several related variables. In problems relating to rates of change, one of these variables will be time (t). Using differentiation, you can find related rates of change.

For example, given $\dfrac{dV}{dt}$ and $\dfrac{dr}{dt}$, look for a relationship between V and r to get:

$$\frac{dV}{dt} = \frac{dV}{dr} \cdot \frac{dr}{dt}$$

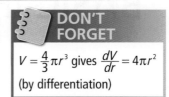

DON'T FORGET

$V = \frac{4}{3}\pi r^3$ gives $\frac{dV}{dr} = 4\pi r^2$

(by differentiation)

Example 2.14

A metal cylinder is rolled in a steel press in such a way that it keeps a cylindrical shape, getting longer and thinner, and its volume, V, remains constant. Before rolling, it has radius 2 cm and length 10 cm. The radius is decreasing at a constant rate of 0·2 cm s^{-1}. At what rate is the length increasing when the radius is 1 cm?

If the radius and length at any time are r and l,

$\dfrac{dr}{dt} = -0\cdot2$ and $V = \pi r^2 l = \pi\, 2^2 10 = 40\pi$, so:

$\pi r^2 l = 40\pi$, hence $l = \dfrac{40}{r^2}$ and:

$$\frac{dl}{dt} = \frac{d}{dt}\left(\frac{40}{r^2}\right) = -\frac{80}{r^3}\frac{dr}{dt} = \frac{16}{r^3}.$$

So, when the radius is 1 cm, the length is increasing at 16 cm s^{-1}.

NOW TRY THIS

(1) Given $y = (x+1)^2(x+2)^{-4}$ and $x > 0$, use logarithmic differentiation to show that $\dfrac{dy}{dx}$ can be expressed in the form $\left(\dfrac{a}{x+1} + \dfrac{b}{x+2}\right)y$, stating the values of the constants a and b. **3**

(2) Obtain the derivatives of the following functions:

   (a) $f(x) = \exp(-\cos 3x)$, **3**

   (b) $y = 2^{x^3+x}$. **3**

(3) Differentiate $f(x) = \sin^{-1}(4x)$, where $-\dfrac{1}{4} < x < \dfrac{1}{4}$. **2**

(4) Given $x = \tan\theta$, $y = \sin\theta$, show that $\dfrac{dy}{dx} = \cos^3\theta$. **3**

(5) A hollow circular cone (see diagram) is fixed in a vertical position and filled at a constant rate of 2 cm^3 per second. Calculate the rate at which the depth, x, is increasing when $x = 3$. **5**

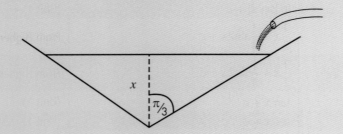

BASIC INTEGRATION

The indefinite integral of a given function $f(x)$, denoted by $\int f(x)dx$, is a function $F(x)$ which satisfies $\frac{dF}{dx} = f$. This notation was introduced by Leibniz in 1675. The $\int$ symbol is an elongated S, meaning 'sum', and arose in the calculation of areas.

The definite integral: $\int_a^b f(x)dx = F(b) - F(a) = \left[F(x)\right]_a^b$

An indefinite integral is a function of x, whereas a definite integral is a **number**.

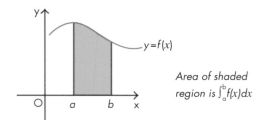

Area of shaded region is $\int_a^b f(x)dx$

DON'T FORGET

$\int x^p dx = \frac{x^{p+1}}{p+1} + c$ for any value of p, including negative and fractional values, **except** the value $p = -1$

Example 3.1   (reminders from Higher Maths)

$\int \frac{1}{x^3}dx = \frac{-1}{2x^2} + c$, where c is an arbitrary constant $\int_1^2 \frac{1}{x^3}dx = \left[\frac{-1}{2x^2}\right]_1^2 = \frac{-1}{8} + \frac{1}{2} = \frac{3}{8}$

INTEGRATION BY SUBSTITUTION

Many integrals can be simplified by replacing x by another carefully chosen variable. A simple case is $u = ax + b$, where a and b are constants, which gives:

$$\int f(ax + b)dx = \frac{1}{a}\int f(u)du.$$

Combining this with the table of standard derivatives (see the table in Chapter 2, page 12) gives the following useful standard integrals:

DON'T FORGET

$\int \frac{1}{x} dx = \ln|x| + c$

$f(x)$	$\int f(x)dx$	
e^x	$e^x + c$	(Unit 1)
$\dfrac{1}{x}$	$\ln|x| + c$ $(or \log_e|x| + c)$	(Unit 1)
$\tan x$	$-\ln|\cos x| + c$ $(or \ln|\sec x| + c)$	(Unit 1)
$\cot x$	$\ln|\sin x| + c$	(Unit 1)
$\sin^2 x$	$\dfrac{1}{2}x - \dfrac{1}{4}\sin 2x + c$	From Higher
$\cos^2 x$	$\dfrac{1}{2}x + \dfrac{1}{4}\sin 2x + c$	From Higher
$\sec^2 x$	$\tan x + c$	(Unit 1)
$\sec x \tan x$	$\sec x + c$	(Unit 1)
$\operatorname{cosec}^2 x$	$-\cot x + c$	(Unit 1)
$\dfrac{1}{\sqrt{1-x^2}},\ \dfrac{1}{\sqrt{a^2-x^2}}$	$\sin^{-1}x + c,\ \dfrac{1}{a}\sin^{-1}\dfrac{x}{a} + c$	(Unit 2)
$\dfrac{1}{1+x^2},\ \dfrac{1}{a^2+x^2}$	$\tan^{-1}x + c,\ \dfrac{1}{a}\tan^{-1}ax + c$	(Unit 2)

contd

INTEGRATION BY SUBSTITUTION   contd

When the substitution is more complicated, you will usually be given the substitution to be used. However, the substitution shown in Example 3.2 is worth remembering.

Example 3.2

$\int \frac{f'(x)}{f(x)} dx$ can be obtained by making the substitution $u = f(x)$. Using the differential notation, $du = f'(x)dx$, giving:

$\int \frac{1}{u} du = \ln|u| + c = \ln|f(x)| + c.$

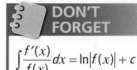

DON'T FORGET

$\int \frac{f'(x)}{f(x)} dx = \ln|f(x)| + c$

Example 3.3

Use the substitution $x = 2 \tan u$ to evaluate $\int_0^2 (4 + x^2)^{-3/2} dx.$

Differentiating the substitution $x = 2 \tan u$ gives:

$dx = 2 \sec^2 u \, du$

when $x = 0$, $u = 0$, and when $x = 2$, $u = \frac{\pi}{4}$.

Now we need to simplify the function to be integrated, so we do this separately before continuing with the integration:

$(4 + x^2)^{-3/2} = (4 \sec^2 u)^{-3/2} = \frac{1}{8 \sec^3 u} = \frac{\cos^3 u}{8}.$

So, we now have:

$\int_0^2 (4 + x^2)^{-3/2} dx = \frac{1}{8} \int_0^{\pi/4} \cos^3 u \, 2 \sec^2 u \, du = \frac{1}{4} \int_0^{\pi/4} \cos u \, du$

$= \frac{1}{4} \left[\sin \frac{\pi}{4} - \sin 0 \right] = \frac{1}{4\sqrt{2}} = \frac{\sqrt{2}}{8}.$

DON'T FORGET

For indefinite integrals, remember to change back to the variable x. For definite integrals, change the limits to correspond to the new variable. Alternatively, change the variable back before evaluating.

Follow these steps when using integration by substitution:

- Obtain dx as (something)du; the something can be left in terms of x here if necessary.
- Change the limits (for a definite integral).
- Substitute for **ALL** occurrences of x and dx.
- Simplify, and make sure the integrand is now completely in terms of u.
- Integrate.
- For an indefinite integral, remember to change back to the variable x.

DON'T FORGET

Where possible and sensible, leave your answer as an exact value.

NOW TRY THIS

(1) Use the substitution $x = (u - 1)^2$ to obtain $\int \frac{1}{(1 + \sqrt{x})^3} dx.$

5

FURTHER INTEGRATION

AREAS AND VOLUMES

Integration can be used to obtain areas under curves, between two curves, and also volumes of revolution.

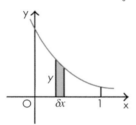

Example 3.4

Calculate the area in the first quadrant bounded by the x-axis, y-axis, and the curve with equation $y = 6 + x - x^2$.

$x^2 - x - 6 = (x + 2)(x - 3)$, so the curve crosses the x-axis at $x = 3$ since in the first quadrant $x > 0$.

Required area $= \int_0^3 (6 + x - x^2)dx = \left[6x + \dfrac{x^2}{2} - \dfrac{x^3}{3}\right]_0^3 = \dfrac{27}{2}$

Example 3.5

A solid is formed by rotating the curve $y = e^{-2x}$ between $x = 0$ and $x = 1$ through 360° about the x-axis. Calculate the volume of the solid.

The volume is given by $\int_0^1 \pi y^2 dx = \pi \int_0^1 e^{-4x}dx = \pi\left[\dfrac{-1}{4}e^{-4x}\right]_0^1 = \dfrac{\pi}{4}\left(1 - e^{-4}\right)$

Volume $\approx \pi y^2 \delta x$

Summing gives $\pi \int_0^1 y^2 dx$

MORE STANDARD INTEGRALS

DON'T FORGET

Make sure you distinguish between:

$\int \dfrac{1}{a^2 + x^2}dx$ and

$\int \dfrac{x}{a^2 + x^2}dx$

$= \dfrac{1}{2}\ln\left(a^2 + x^2\right) + c$

Here are two more standard integrals which you need to know:

$\int \dfrac{1}{\sqrt{a^2 - x^2}}dx = \sin^{-1}\left(\dfrac{x}{a}\right) + c$

$\int \dfrac{1}{a^2 + x^2}dx = \dfrac{1}{a}\tan^{-1}\left(\dfrac{x}{a}\right) + c$

These can be remembered, or are easily obtained using the substitutions $x = a \sin u$ and $x = a \tan u$ respectively. Alternatively, the substitution $x = at$ may be used with both integrals.

Example 3.6

$\int_0^2 \dfrac{1}{x^2 + 4}dx = \dfrac{1}{2}\left[\tan^{-1}\left(\dfrac{x}{2}\right)\right]_0^2 = \dfrac{1}{2}\tan^{-1}(1) = \dfrac{\pi}{8}$

DON'T FORGET

Make sure you distinguish between:

$\int \dfrac{1}{\sqrt{a^2 - x^2}}dx$ and

$\int \dfrac{x}{\sqrt{a^2 - x^2}}dx$

$= -\sqrt{a^2 - x^2} + c$

NOW TRY THIS

(1) (a) Use the substitution $u = 1 + x^2$ to obtain $\int_0^1 \dfrac{x^3}{(1+x^2)^4}dx$.   5

     (b) A solid is formed by rotating the curve $y = \dfrac{x^{3/2}}{(1+x^2)^2}$ between $x = 0$ and $x = 1$

     through 360° about the x-axis. Write down the volume of this solid.   1

(2) Use the substitution $x = 2 \sin \theta$ to obtain the exact value of $\int_0^{\sqrt{2}} \dfrac{x^2}{\sqrt{4-x^2}}dx$.   6

     (Note that $\cos 2A = 1 - 2\sin^2 A$.)

Many integrals require additional techniques to reduce them to standard forms. We have already met the method of substitution, and we now look at **partial fractions** and **integration by parts**.

PARTIAL FRACTIONS

Integrals of the form $\int \frac{p(x)}{q(x)} dx$, where p and q are polynomials, can be obtained by expressing $\frac{p(x)}{q(x)}$ in partial fractions. See Chapter 1 for details about partial fractions.

Example 3.7

Obtain $\int \frac{x^2}{x^2 - x - 2} dx$.

Using long division gives: $\frac{x^2}{x^2 - x - 2} = 1 + \frac{x+2}{x^2 - x - 2}$.

We now set $\frac{x+2}{x^2 - x - 2} = \frac{A}{x+1} + \frac{B}{x-2}$ for constants A and B.

This gives $x + 2 = A(x - 2) + B(x + 1)$, from which we get $A = -\frac{1}{3}$, $B = \frac{4}{3}$.

Hence $\int \frac{x^2}{x^2 - x - 2} dx = \int \left(1 + \frac{4/3}{x-2} - \frac{1/3}{x+1} \right) dx = x + \frac{4}{3}\ln|x-2| - \frac{1}{3}\ln|x+1| + c$.

Example 3.8

Evaluate $\int_0^2 \frac{x-1}{(x-3)^2} dx$.

Let $\frac{x-1}{(x-3)^2} = \frac{A}{x-3} + \frac{B}{(x-3)^2}$.

This gives $x - 1 = A(x - 3) + B$ and hence $A = 1$ and $B = 2$.

Applying this result gives $\int_0^2 \frac{x-1}{(x-3)^2} dx = \int_0^2 \left(\frac{1}{x-3} + \frac{2}{(x-3)^2} \right) dx$

$$= \left[\ln|x-3| - \frac{2}{x-3} \right]_0^2 = \frac{4}{3} - \ln 3.$$

Example 3.9

Obtain $\int \frac{1}{x^3 + 4x} dx$.

We factorise $x^3 + 4x$ to get $x(x^2 + 4)$ and let $\frac{1}{x^3 + 4x} = \frac{A}{x} + \frac{Bx + C}{x^2 + 4}$.

So, $A(x^2 + 4) + x(Bx + C) = 1$.

Setting $x = 0$ gives $A = ¼$.

Equating coefficients of x^2 gives: $A + B = 0 \Rightarrow B = -¼$.

Equating coefficients of x gives $C = 0$.

Then $\int \frac{1}{x^3 + 4x} dx = \frac{1}{4} \int \left(\frac{1}{x} - \frac{x}{x^2 + 4} \right) dx = \frac{1}{4}\ln|x| - \frac{1}{8}\ln(x^2 + 4) + c$.

> **DON'T FORGET**
> Partial fractions can be useful for differentiation as well as integration.

> **DON'T FORGET**
> Remember the modulus sign in log functions.

> **DON'T FORGET**
> Note that the modulus sign is absolutely essential here, and its omission would lose marks.

> **DON'T FORGET**
> The form of the polynomial q in $\frac{p(x)}{q(x)}$ could be any of the following types:
> - a quadratic function
> - a product of three linear factors
> - the product of a linear factor and an irreducible quadratic factor

NOW TRY THIS

(1) (a) Express $\frac{1}{x^2 + 2x - 8}$ in partial fractions.   **2**

    (b) Evaluate $\int_0^1 \frac{1}{x^2 + 2x - 8} dx$.   **4**

(2) (a) Express $\frac{1}{x^3 + x}$ in partial fractions.   **4**

    (b) Obtain a formula for $I(k)$, where $I(k) = \int_1^k \frac{1}{x^3 + x} dx$, expressing it in the form $\ln\left(\frac{a}{b}\right)$, where a and b depend on k.   **4**

INTEGRATION BY PARTS

Integration by parts is the integration equivalent of the product rule from differentiation. The idea of integration by parts is to produce an integral which is easier to deal with. If you are left with an integral that looks harder than, or as hard as, the integral you started with, then you may not have made the best choice of f and g'. Swapping them over may be all that is required.

The basic result is $\int f(x)g'(x)dx = f(x)g(x) - \int f'(x)g(x)dx$, or the equivalent form:

$\int u\dfrac{dv}{dx}dx = uv - \int v\dfrac{du}{dx}dx$ (with a corresponding result for definite integrals).

Its application requires a sensible choice of the functions f and g (or u and v).

The mnemonic LIPET can be useful in making your choice of function for u (the one you are going to differentiate).

   L = log

   I = inverse trig

   P = polynomial

   E = exponential

   T = trig

Choose u or $f(x)$ in this order.

Example 3.10

Evaluate $\displaystyle\int_{0}^{2}(x+1)(x-2)^3dx$. Let $f(x) = x + 1$, $g'(x) = (x - 2)^3$ to give:

$$\int_{0}^{2}(x+1)(x-2)^3dx = \left[(x+1)\frac{(x-2)^4}{4}\right]_{0}^{2} - \frac{1}{4}\int_{0}^{2}(x-2)^4dx$$

$$= -4 - \frac{1}{20}\left[(x-2)^5\right]_{0}^{2} = -\frac{28}{5}.$$

Note that, in the above example, the alternative choice of $g'(x) = x + 1$ and $f(x) = (x - 2)^3$ leads to an integral which requires several integrations by parts.

Example 3.11   Repeated application

DON'T FORGET

Sometimes the method may need a second application.

Obtain $\int x^2 \sin 2x \ dx$.

$f(x) = x^2 \qquad\qquad g(x) = -\dfrac{1}{2}\cos 2x$

$\downarrow \qquad\qquad\qquad \uparrow$

$f'(x) = 2x \qquad g'(x) = \sin 2x$

A first application gives $\displaystyle\int x^2 \sin 2x \, dx = -\frac{x^2}{2}\cos 2x + \int x\cos 2x \, dx$.

In this second integral, we set $f(x) = x$, $g'(x) = \cos 2x$, which gives:

$$\int x\cos 2x \, dx = \frac{x}{2}\sin 2x - \int\frac{\sin 2x}{2}dx = \frac{x}{2}\sin 2x + \frac{1}{4}\cos 2x + c.$$

Combining these results, we have:

$$\int x^2 \sin 2x \, dx = -\frac{x^2}{2}\cos 2x + \frac{x}{2}\sin 2x + \frac{1}{4}\cos 2x + c.$$

For some functions f, $\int f(x)dx$ can be obtained by setting $u = f(x)$, $\dfrac{dv}{dx} = 1$.

Example 3.12

Evaluate $\displaystyle\int_0^{1/2} \sin^{-1}2x\,dx$.

Set $u = \sin^{-1} 2x$ and $\dfrac{dv}{dx} = 1$, so:

$\dfrac{du}{dx} = \dfrac{2}{\sqrt{1-4x^2}}$ and $v = x$.

Thus $\displaystyle\int_0^{1/2} \sin^{-1}2x\,dx = \Big[x\sin^{-1}2x \Big]_0^{1/2} - \int_0^{1/2} \dfrac{2x}{\sqrt{1-4x^2}}\,dx$.

The first term is $\dfrac{1}{2}\sin^{-1}1 = \dfrac{\pi}{4}$.

The second integral can be obtained in various ways.

If you recognise that the numerator is the derivative of $1 - 4x^2$, apart from a constant multiplicative factor, you can guess that the integral must be of the form $k\sqrt{1-4x^2}$, and it is then easy to decide what k must be.

Alternatively, make the substitution $u = 4x^2$.

Either method leads to the result:

$$\int_0^{1/2} \dfrac{2x}{\sqrt{1-4x^2}}\,dx = \left[-\dfrac{1}{2}\sqrt{1-4x^2} \right]_0^{1/2} = \dfrac{1}{2}.$$

Combining these, we get $\displaystyle\int_0^{1/2} \sin^{-1}2x\,dx = \dfrac{\pi}{4} - \dfrac{1}{2}$.

⚙ NOW TRY THIS

(1) Use integration by parts to obtain $\displaystyle\int 10x^2 \cos 5x\,dx$.   **5**

(2) Use integration by parts to obtain the exact value of $\displaystyle\int_0^1 x\tan^{-1}x^2\,dx$.   **5**

(3) Use integration by parts to obtain $\displaystyle\int \dfrac{(x+3)^2}{(x+1)^3}\,dx$.   **5**

(4) Let $I = \displaystyle\int_0^{\pi/2} e^{2x}\cos x\,dx$ and $J = \displaystyle\int_0^{\pi/2} e^{2x}\sin x\,dx$.

Use integration by parts to show that $I = e^{\pi} - 2J$, and also that $J = 1 + 2I$.

Hence show that $\displaystyle\int_0^{\pi/2} e^{2x}\cos x\,dx = \dfrac{1}{5}(e^{\pi} - 2)$.   **6**

PROPERTIES OF FUNCTIONS

A function f requires **two** things to specify it completely:

- a **domain**, which is a set of values x on which the function is defined
- a **rule** which, for any value x in the domain, determines a **unique** value y (y need not belong to the domain).

We write $y = f(x)$, and say that y is the value of f at x. The set of all values y obtained this way is called the **range** of f.

Example 4.1

$f(x) = -\sqrt{x}$ for all $x > 0$.

Here the domain is the set of all positive real numbers, while the range is the set of all negative real numbers.

In Advanced Higher, the rule is usually given by a formula, or combination of formulae. If the domain is not specified, it can be assumed to be $\mathbb{R}$ (set of all real numbers), or the largest subset of $\mathbb{R}$ for which the rule makes sense. The range is not usually specified, but you could be asked to obtain it.

EVEN AND ODD FUNCTIONS

If $f(x) = f(-x)$ for all x in the domain of f, then f is said to be **even**. The graph of f is symmetrical, with $x = 0$ the axis of symmetry.

If $f(x) = -f(-x)$ for all x in the domain of f, then f is said to be **odd**. The graph of f has **half-turn** symmetry about the origin.

For polynomial functions:

All ODD powers $\rightarrow$ ODD function: e.g. $x^3 + 2x$.

All EVEN powers $\rightarrow$ EVEN function: e.g. $x^6 - 2x^2 + 3$.

Example 4.2

$f(x) = \cos x + x^2$: f is even.

$g(x) = x + \sin x$: g is odd.

$h(x) = x^3 + x^2 + x$: h is neither even nor odd, but it is the sum of an even function (the function x^2) and an odd function (the function $x^3 + x$).

INVERSE FUNCTIONS

For some functions, f, each value, y, in the range comes from just one value of x. In this case, the equation $y = f(x)$ can be solved to give x as a function of y. This function is called the **inverse function** of f, denoted by f^{-1}. The domain of f^{-1} is the range of f, and its range is the domain of f.

From the definition of f^{-1}, it follows that $f(f^{-1}(x)) = x = f^{-1}(f(x))$, for any function f which has an inverse.

Example 4.3

$f(x) = x^2$, all x in $\mathbb{R}$.

For any given value of y, with $y \geqslant 0$, the equation $f(x) = x^2$ is a quadratic equation for x, giving two values of x, so f does not have an inverse.

However, $g(x) = x^2$, $x \geqslant 0$, **does** have an inverse.

DON'T FORGET

Do not confuse this use of the terms 'even' and 'odd' with even and odd integers.

DON'T FORGET

ODD function + ODD function = ODD function.

DON'T FORGET

EVEN function + EVEN function = EVEN function.

DON'T FORGET

Constant functions are even.

DON'T FORGET

The notation $f^{-1}(x)$ is not to be confused with a negative index: $f^{-1}(x)$ does **not** mean $\frac{1}{f(x)}$.

Note: not every function has an inverse.

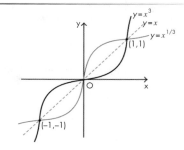

FEATURES OF GRAPHS

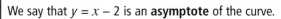

If $y = f(x)$, and we plot the points (x, y), we get the graph of f.

If the function f has an inverse, every horizontal line meets the graph $y = f(x)$ once at most. The graph of f^{-1} is obtained by reflecting the graph of f in the line $y = x$.

Asymptotes

Consider the curve $y = x - 2 + \dfrac{1}{x^2 + 1}$ where x is real.

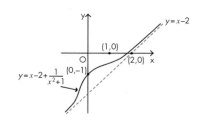

As x gets large, $x^2 + 1$ gets very large and $\dfrac{1}{x^2 + 1}$ gets very small.

So, the value of y gets closer and closer to the value $x - 2$, which means that, for large x, points on the curve are close to the line $y = x - 2$.

We say that $y = x - 2$ is an **asymptote** of the curve.

Some graphs have vertical asymptotes. If $f(a)$ is not defined, and $f(x)$ becomes larger and larger as $x \to a$, then we say that $x = a$ is a vertical asymptote of the curve $y = f(x)$.

Example 4.4

The curve defined by the equation $y = 1 + \dfrac{1}{x + 2}$ has two

asymptotes: the horizontal asymptote $y = 1$ and the vertical asymptote $x = -2$.

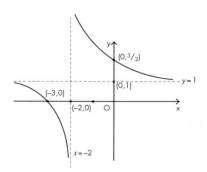

The details of the diagram can be confirmed after reading the next section.

Most functions met at Advanced Higher are continuous, except at isolated points such as vertical asymptotes. An intuitive idea of continuity is that the graph of a continuous function has no breaks in it.

NOW TRY THIS

(1) Determine whether the function $f(x) = x^3 \tan x$ is odd, even or neither. Justify your answer.   **3**

(2) The diagram shows part of the graph of a function f which satisfies the following conditions:

   (i) f is an even function;

   (ii) two of the asymptotes of the graph $y = f(x)$ are $y = x$ and $x = 1$.

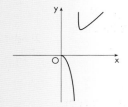

   Copy the diagram and complete the graph.
   Write down equations for the other two asymptotes.   **3**

(3) Part of the graph $y = f(x)$ is shown alongside, where the dotted lines indicate asymptotes. Sketch the graph $y = -f(x + 1)$, showing its asymptotes. Write down the equations of the asymptotes.   **4**

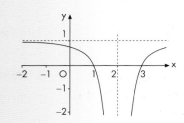

(4) The diagram below shows part of the graph of a function $f(x)$.
State whether $f(x)$ is odd, even or neither. Fully justify your answer.   **3**

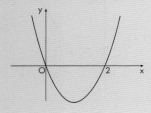

CRITICAL POINTS

http://tutorial.math.lamar.edu/Classes/Calci/DerivAppsIntro.aspx

Any point $(a, f(a))$ on the graph of f for which either $f'(a) = 0$ or $f'(a)$ does not exist is called a **critical point** of f.

If $f'(a) = 0$, the point $(a, f(a))$ is called a **stationary** point.

Example 4.5

Consider the function defined for $-1 \leqslant x \leqslant 1$ by $f(x) = |x|$. The graph of f shows that f is continuous; $f'(0)$ does not exist, since it changes **instantaneously** from a value of -1 when $x < 0$ to a value of $+1$ when $x > 0$. (In other words, the change in $f'(x)$ value is not 'smooth' – visually, there is a 'kink' at the origin.) So, $(0,0)$ is a critical point. The points $(-1, 1)$ and $(1, 1)$ are also critical points, because $x = -1$ and $x = 1$ are the end points of this function.

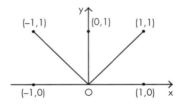

A point (x_0, y_0) on the graph $y = f(x)$ where the curve changes from concave up, when $f''(x) > 0$, to concave down, when $f''(x) < 0$, or vice versa, is called a **point of inflexion**. At such a point, $f''(x_0) = 0$.

Testing for a point of inflexion requires that you find the x-value for which $f''(x) = 0$ and check that the value of $f''(x)$ changes sign to the left and right of this x-value.

1. Rising stationary point of inflexion

$f'(a) = 0$ AND $f''(a) = 0$

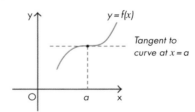

2. Falling stationary point of inflexion

$f'(a) = 0$ AND $f''(a) = 0$

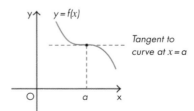

3. Sloping increasing point of inflexion

$f'(a) > 0$ AND $f''(a) = 0$

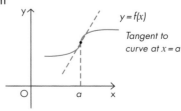

4. Sloping decreasing point of inflexion

$f'(a) < 0$ AND $f''(a) = 0$

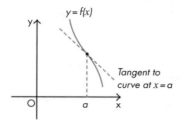

Thus, the gradient need **not** be 0 at a point of Inflexion.

Example 4.6

Consider $f(x) = x^3$ for real x.

$f''(x) = 6x$, so $(0,0)$ is a point of inflexion.

The curve changes from concave down for $x < 0$ to concave up for $x > 0$.

Critical points given by solving $f'(x) = 0$ fall into one of three groups:

- local or global **maxima**

- local or global **minima**

- **points of inflexion**.

Maxima and minima can be determined by looking at the sign of f' near the critical point (nature table). Alternatively, if f'' is easy to obtain, look at the sign of f'' at the critical point.

- $f''(a) > 0 \implies$ minimum at $x = a$

- $f''(a) < 0 \implies$ maximum at $x = a$

- $f''(a) = 0 \implies$ stationary point or a point of inflexion.

$f''(x) > 0$ means curve is concave up, so if $f'(a) = 0$ and $f''(a) > 0$ we have

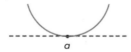

$f''(x) < 0$ means curve is concave down, so if $f'(a) = 0$ and $f''(a) < 0$ we have

$f'(a) = 0 = f''(a)$ can be a point of inflexion or a stationary point.

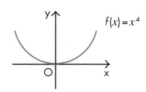

For $f(x) = x^4$, $f'(0) = f''(0)$ with minimum at $x = 0$.

DON'T FORGET

Stationary points may be local maxima/minima, or points of inflexion.

DON'T FORGET

A point of inflexion need not be a stationary point.

DON'T FORGET

If the domain of a function is a finite interval, then the end points are always critical points.

contd

Example 4.7

$f(x) = \frac{x^2 - 1}{x^2 + 1}$.

Dividing gives $f(x) = 1 - \frac{2}{x^2 + 1}$, which makes it easier to obtain f'.

$f'(x) = \frac{4x}{(x^2 + 1)^2}$ and $f''(x) = \frac{4 - 12x^2}{(x^2 + 1)^3}$.

Note that the graph $y = f(x)$ is symmetrical about the y-axis because f is even: $(f(x) = f(-x))$.

There are no vertical asymptotes since $x^2 + 1 \neq 0$, but because $f(x) \to 1$ as $x \to \pm\infty$, $y = 1$ is a horizontal asymptote.

Stationary points are given by $f'(x) = 0 \Rightarrow x = 0$, so $(0, -1)$ is a stationary point. As $f''(0) > 0$, this gives a global minimum. Alternatively, you can easily check that the gradient as x passes through $x = 0$ changes: ↘ ― ↗.

To look for the existence of points of inflexion, we solve $f''(x) = 0$ to find that $x = \pm^1/_{\sqrt{3}}$. In this case, we only need to investigate what happens for the positive value of x, because the graph is symmetrical about the y-axis.

$f''\left(\frac{1}{\sqrt{3}}\right) = 0$ but $f'\left(\frac{1}{\sqrt{3}}\right) \neq 0$, so this gives a sloping point of inflexion.

To determine whether this is a rising/increasing point of inflexion or falling/decreasing point of inflexion, use either:

* nature table for $f(x)$

* nature table for $f''(x)$ (i.e. concavity of curve).

* $f'\left(\frac{1}{\sqrt{3}}\right) > 0 \Rightarrow$ rising point of inflexion

* $f'\left(\frac{1}{\sqrt{3}}\right) < 0 \Rightarrow$ falling point of inflexion.

The curve changes from concave up to concave down

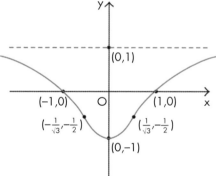

Example 4.8

$f(x) = x^4 - 4x^3$

$f'(x) = 4x^3 - 12x^2$ and $f''(x) = 12x^2 - 24x$

$f'(x) = 0 \Rightarrow x = 0$ and $x = 3$.

Using a nature table for f', we get:

	0		3	
→		→		→
−	0	−	0	+
↘	−	↘	−	↗

so there is a horizontal point of inflexion at $x = 0$, and a (global) minimum at $x = 3$.

The position of points of inflexion are given by solving $f''(x) = 12x^2 - 24x = 0$, giving $x = 0$ (already obtained) and $x = 2$. As $f''(x) < 0$ for $x < 2$ and $f''(x) > 0$ for $x > 2$, the curve changes from concave down to concave up at $x = 2$.

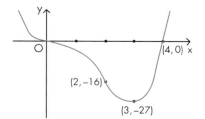

DON'T FORGET

Symmetry
Axes
Behaviour at $\pm\infty$
Undefined
Critical Points

Note that in this example the 2nd derivative test for the nature of any turning points is not completely successful.

Because $f'(3) = 0$ and $f''(3) > 0$, we can conclude that $x = 3$ gives a minimum, as above, but because $f'(0) = f''(0) = 0$ we have to resort to a nature table for $x = 0$.

DON'T FORGET

Before differentiating, consider the possibility of simplifying $f(x)$, for example by using partial fractions or division.

Steps for sketching functions

- Is there any obvious symmetry (odd, even, neither)?
- Find, if possible, where the curve cuts the x- and y-axes.
- Examine the behaviour of the function as $x \to \pm\infty$, including horizontal and oblique asymptotes.
- Investigate critical points.
- Investigate any values of x for which the function is undefined (vertical asymptotes).

NOW TRY THIS

(1) The function f is defined by $f(x) = \dfrac{x^2 + 3x}{x + 1}$ $\quad (x \neq -1)$.

    (a) Obtain equations of the asymptotes of the graph of f.   **3**

    (b) Show that the graph is always increasing.   **3**

    (c) Sketch the graph of f, showing all important features.   **2**

(2) The function f is defined by $f(x) = \dfrac{x - 3}{x + 2}$, $x \neq -2$, and the diagram shows part of its graph.

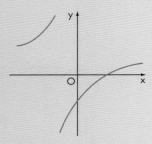

    (a) Obtain algebraically the asymptotes of the graph of f.   **3**

    (b) Prove that f has no stationary values.   **2**

    (c) Does the graph of f have any points of inflexion? Justify your answer.   **2**

    (d) Sketch the graph of the inverse function, f^{-1}. State the asymptotes and the domain of f^{-1}.   **3**

RELATED GRAPHS

DON'T FORGET

The graph of $f(x + k)$ is obtained by shifting the graph of $f(x)$ k units along the x-axis to the **left** when $k > 0$.

If you are given a graph $y = f(x)$ (sometimes only as a diagram without a formula for f), there are several related graphs which you could be asked to sketch. In what follows, k is a constant.

(i) $y = kf(x)$, (ii) $y = f(kx)$, (iii) $y = f(x) + k$, (iv) $y = f(x + k)$, (v) $y = |f(x)|$, (vi) $y = f'(x)$, (vii) a combination of (i) to (vi).

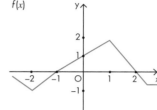

Example 4.9

The diagram shows the graph of a function f.

Sketch the graphs of $2f(x)$, $f(2x)$, $f(x) + 2$, $f(x + 2)$, $|f(x)|$.

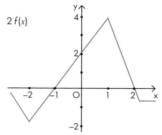

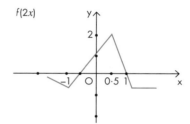

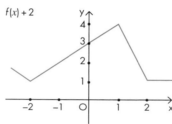

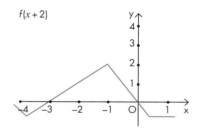

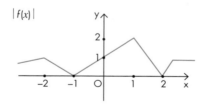

NOW TRY THIS

(1)

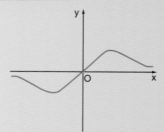

The diagram shows the shape of the graph of $y = \dfrac{x}{1 + x^2}$. Obtain the stationary points of the graph. 4

Sketch the graph of $y = \left|\dfrac{x}{1 + x^2}\right|$ and identify its three critical points. 3

GRAPHS OF TRIGONOMETRIC FUNCTIONS

You should be able to sketch the graphs of the three basic functions $\sin x$, $\cos x$ and $\tan x$, their inverse functions and simple related functions, e.g. $k \cos x$ and $\sin kx$.

$\sin x$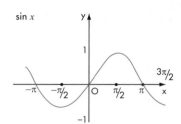

$\cos x$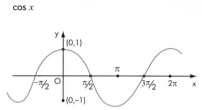

$\tan x$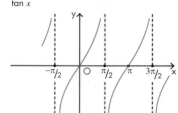

$\sin^{-1} x$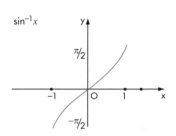

$\cos^{-1} x$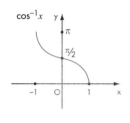

$\tan^{-1} x$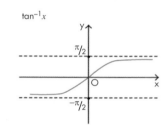

Example 4.10

Sketch the graph of $\sec x$ over the interval $-\pi \leqslant x \leqslant \pi$.

$\sec x$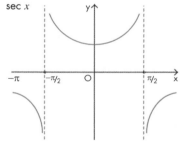

There are three things to note about the graph of $\sec x$ before you start:

1 The range is in two parts, $y \geqslant 1$ and $y \leqslant -1$.

2 There are vertical asymptotes at $x = \pm\dfrac{\pi}{2}$.

3 The graph is symmetrical about $x = 0$.

To define the inverse function, we must restrict the domain to $0 \leqslant x \leqslant \pi$, which gives the graph of $\sec^{-1} x$ as shown in the graph here:

$\sec^{-1} x$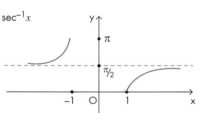

NOW TRY THIS

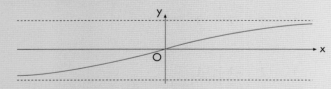

(1) (a) The diagram shows part of the graph of $f(x) = \tan^{-1} 2x$ and its asymptotes. State the equations of these asymptotes. **2**

(b) Use integration by parts to find the area between $f(x)$, the x-axis and the lines $x = 0$, $x = \dfrac{1}{2}$. **5**

(c) Sketch the graph of $y = |f(x)|$ and calculate the area between this graph, the x-axis and the lines $x = -\dfrac{1}{2}$, $x = \dfrac{1}{2}$. **3**

SYSTEMS OF LINEAR EQUATIONS AND GAUSSIAN ELIMINATION

 http://tutorial.math.lamar.edu/Classes/Alg/Systems.aspx

Linear equations only have variables to the power of 1 and no products of variables. Many problems reduce to solving a system of linear equations; and there is a systematic way of doing this.

Example 5.1

Consider the following system of equations:

$$2x + y + 3z = 6, \quad x + 2y - z = -1, \quad x + y + 2z = 3.$$

There are three main steps to solving this system.

Step 1: Write the system of equations as an **augmented matrix**. The numbers (elements) in the first column of this matrix are the coefficients of x, those in the second column the coefficients of y, and so on.

$$\begin{array}{ccc|c} 2 & 1 & 3 & 6 \\ 1 & 2 & -1 & -1 \\ 1 & 1 & 2 & 3 \end{array}$$

We now describe a set of operations, called **elementary row operations**, which replace this matrix by a simpler matrix from which we can easily obtain the solution of our system of linear equations.

When we multiply, for example, the 3rd row of the matrix by 2, this corresponds to the equation $2x + 2y + 4z = 6$, which is equivalent to the 3rd equation.

 **DON'T FORGET**

There are three elementary row operations:

- any row of the matrix can be multiplied by any (non-zero) number
- any row can be changed by adding any multiple of another row to it
- any two rows can be interchanged.

The object is to reduce the initial matrix to upper (or lower) triangular form. This means producing zeros in the bottom left-hand triangle in the upper triangular form (column 1 rows 2 and 3, and column 2 row 3). For the lower triangular form, the zeros are in row 1 columns 2 and 3, and row 2 column 3.

Step 2: Simplify the augmented matrix using elementary row operations.

$$\begin{array}{ccc|c} 2 & 1 & 3 & 6 \\ 1 & 2 & -1 & -1 \\ 1 & 1 & 2 & 3 \end{array} \rightarrow$$

$$\begin{array}{ccc|c} 1 & 2 & -1 & -1 \\ 2 & 1 & 3 & 6 \\ 1 & 1 & 2 & 3 \end{array}$$

(interchanging rows 1 and 2)

If possible, start with a 1 in the top left corner of the augmented matrix. Interchanging two rows may do this. It will avoid the use of fractions to begin with.

We now subtract $2 \times$ row 1 from row 2, for which we use the shorthand notation: $R_2' = R_2 - 2R_1$, and also subtract row 1 from row 3, $R_3' = R_3 - R_1$:

$$\begin{array}{ccc|cl} 1 & 2 & -1 & -1 & \\ 0 & -3 & 5 & 8 & R_2 - 2R_1 \\ 0 & -1 & 3 & 4 & R_3 - R_1 \end{array}$$

The next step is to get a zero in row 3 column 2. We accomplish this by $R_3' = 3R_3 - R_2$, where R_2 and R_3 now refer to the new version of the original augmented matrix. (Note that there are a number of alternative acceptable shorthand notations.)

$$\begin{array}{ccc|cl} 1 & 2 & -1 & -1 & \\ 0 & -3 & 5 & 8 & \\ 0 & 0 & 4 & 4 & 3R_3 - R_2 \end{array}$$

This last matrix is in **upper triangular form**, the simplified form required.

Step 3: Solve the system of linear equations.

The system given initially:

$$2x + y + 3z = 6, \quad x + 2y - z = -1, \quad x + y + 2z = 3$$

is equivalent to the system obtained from the final matrix:

$$x + 2y - z = -1, \quad -3y + 5z = 8, \quad 4z = 4.$$

This is because every row operation on the augmented matrix corresponds to an equivalent operation on the system of equations **which does not affect the solution**. This process is familiar to you from previous years' work on simultaneous equations involving only 2 variables.

Working backwards (back substitution) with the simplified system, we have:

$$z = 1, \quad -3y + 5 = 8, \quad \text{so } y = -1, \quad \text{and } x - 2 - 1 = -1, \quad \text{so } x = 2.$$

The method of reducing a system of equations using elementary row operations, followed by back substitution, is known as **Gaussian elimination**, after the German mathematician Karl Friedrich Gauss (1777–1855), widely regarded as one of the greatest mathematicians ever. Although the technique was known to Chinese mathematicians nearly 2000 years prior to Gauss' time, he established its secure basis within a general theory of equation-solving.

DON'T FORGET

Interchange rows if it will give a 1 in the top left corner of the matrix.

DON'T FORGET

1. Augmented matrix
2. Simplify
3. Solve / interpret

NOW TRY THIS

(1) Use Gaussian elimination to obtain solutions of the equations:

$$2x - y + 2z = 1$$
$$x + y - 2z = 2$$
$$x - 2y + 4z = -1.$$

5

(2) Use Gaussian elimination to solve the following system of equations:

$$x + y - z = 6$$
$$2x - 3y + 2z = 2$$
$$-5x + 2y - 4z = 1.$$

5

SPECIAL CASES

REDUNDANCY

Redundancy means that one of the equations can be derived from the others, so adds no further information. It arises when a plane, given by one of the equations, contains the line of intersection of two other planes, given by the other two equations.

Example 5.2

Solve the system of equations $x - 2y + z = 1$, $3x + y - z = 0$, $5x - 3y + z = 2$.

$$
\begin{array}{ccc|c}
1 & -2 & 1 & 1 \\
3 & 1 & -1 & 0 \\
5 & -3 & 1 & 2
\end{array}
\qquad
\begin{array}{ccc|cl}
1 & -2 & 1 & 1 & \\
0 & 7 & -4 & -3 & R_2 - 3R_1 \\
0 & 7 & -4 & -3 & R_3 - 5R_1
\end{array}
$$

Note that the second and third rows are the same, which leads to:

$$
\begin{array}{ccc|cl}
1 & -2 & 1 & 1 & \\
0 & 7 & -4 & -3 & \\
0 & 0 & 0 & 0 & R_3 - R_2
\end{array}
$$

Because there is no equation for z, **z can have any value**. Set $z = t$ to emphasise this, where t is a parameter which can take any value in $\mathbb{R}$.

Solving for y then x gives $z = t$, $y = -\dfrac{3}{7} + \dfrac{4}{7}t$, $x = \dfrac{1}{7} + \dfrac{t}{7}$.

These equations give points that all lie on a straight line (see Chapter 9).

The system of equations has an infinite number of solutions. There are, really, only two equations, and the third is redundant.

INCONSISTENCY

Inconsistency means that there are no values of x, y, z which satisfy all three equations. It arises when three planes, given by the three equations, do not have a point in common.

DON'T FORGET

A system of three linear equations for three unknowns satisfies one of the following:

- There is a unique solution
- There are infinitely many solutions (redundancy)
- There are no solutions (inconsistency).

Example 5.3

Solve the system of equations $x - y - 2z = 0$, $3x + 2y - z = 2$, $2x + 3y + z = 1$.

$$
\begin{array}{ccc|c}
1 & -1 & -2 & 0 \\
3 & 2 & -1 & 2 \\
2 & 3 & 1 & 1
\end{array}
\qquad
\begin{array}{ccc|cl}
1 & -1 & -2 & 0 & \\
0 & 5 & 5 & 2 & R_2 - 3R_1 \\
0 & 5 & 5 & 1 & R_3 - 2R_1
\end{array}
$$

$$
\begin{array}{ccc|cl}
1 & -1 & -2 & 0 & \\
0 & 5 & 5 & 2 & \\
0 & 0 & 0 & 1 & R_2 - R_3
\end{array}
$$

This last third row says that $0x + 0y + 0z = 1$, i.e. $0 = 1$, which is clearly impossible.

The system of equations has no solution, and is said to be inconsistent.

Geometrical interpretations of redundancy and inconsistency in terms of the intersection of three planes are given in Chapter 9.

ILL-CONDITIONING

All our calculations so far have been exact. However, in many real-life problems using experimental or collected data, the coefficients in the equations are often rounded to the nearest integer, or one decimal place, and so on. We will look at a couple of examples where some of the coefficients have been rounded to the nearest integer.

Example 5.4

The equations $2x + y = 1$ and $2x - y = 2$ have the solution $x = 0.75, y = -0.5$.

Suppose that the coefficients of x have been rounded and that the equations should be: $1.9x + y = 1, 1.9x - y = 2$. The solution is now $x = 0.8, y = -0.5$, to 1 decimal place.

So, a 5% change in some of the coefficients has led to about a 5% change in the answer. A situation like this in a real-life problem would be satisfactory.

Example 5.5

The equations $2x - y = -5$ $(20x - 10y = -50)$ and $21x - 10y = -10$ have the solution $x = 40$ and $y = 85$.

Now suppose that the equations should be: $19x - 10y = -50$ and $22x - 10y = -10$.

The solution now is $x = 13\frac{1}{3}$ and $y = 30\frac{1}{3}$.

This time, a 5% change in some of the coefficients has led to a 67% change in x and a 64% change in y. This set of equations is said to be **ill-conditioned**. A small percentage change in the coefficients has led to a significantly larger percentage change in the solution. This set of equations would not therefore form a good model for real-life contexts.

Both of these examples involve calculating the point of intersection of a pair of straight lines. In Example 5.4, the angle between the lines is large (their gradients are 2 and –2). However, in Example 5.5, the two lines are nearly parallel (their gradients are 2 and 2.1). A small change in any of the coefficients in the equations represents a small change in either line; and, because they are nearly parallel, this produces a big change in the point of intersection.

A similar situation holds for some systems of linear equations in three unknowns, where now the solution represents the intersection of planes.

NOW TRY THIS

(1) (a) Use elementary row operations to reduce the following system of equations to upper triangular form:

$x + y + 3z = 1$ $3x + ay + z = 0$ $x + y + z = -1$.   2

(b) Hence express x, y and z in terms of the parameter a.   2

(c) Explain what happens when $a = 3$.   2

(2) (a) Use Gaussian elimination to solve the system of equations below when $\lambda \neq 2$:

$x + y + 2z = 1$ $2x + \lambda y + z = 0$ $3x + 3y + 9z = 5$.   4

(b) Explain what happens when $\lambda = 2$.   2

(3) The set of equations $11y + 7x = 19$, $3y + 2x = 5$ have solution $x = -2, y = 3$.

Obtain the solution to: $11y + 7.3x = 19$, $3y + 1.9x = 5$

and comment on your results.   4

http://tutorial.math.lamar.edu/Classes/LinAlg/MatrixArithmetic.aspx

DON'T FORGET

Order: row by column.

An array of numbers (or letters) contained in brackets is called a **matrix**. The numbers (or letters) in the matrix are called its **elements**.

Example 6.1

$$A = \begin{pmatrix} 1 & 3 \\ 2 & 1 \end{pmatrix}, \quad B = \begin{pmatrix} 5 & 1 & 3 \\ -1 & 0 & x \end{pmatrix}, \quad C = \begin{pmatrix} 2 \\ 1 \\ 1 \end{pmatrix} \text{ are matrices.}$$

A is a 2×2 matrix (read as *two by two*), and is also a **square matrix**. It has **order** 2×2.

B has order 2×3, having two rows and three columns.

C is a column matrix.

Matrices are particularly useful when we define algebraic operations on them.

Two matrices are said to be equal when they have the same order and all corresponding elements are equal.

Example 6.2

The matrices $\begin{pmatrix} 1 & 3 \\ 2 & 1 \end{pmatrix}$ and $\begin{pmatrix} 3 & 1 \\ 1 & 2 \end{pmatrix}$ are not equal. $\begin{pmatrix} 1 & 3 \\ 2 & 1 \end{pmatrix}$ and $\begin{pmatrix} 1 & 3 \\ x & 1 \end{pmatrix}$ are only equal when $x = 2$.

BASIC MATRIX OPERATIONS

ADDITION OF MATRICES

Only matrices of the same order can be added. Simply add corresponding elements.

Example 6.3

$$\begin{pmatrix} 1 & 3 \\ 2 & 1 \end{pmatrix} + \begin{pmatrix} 2 & -1 \\ 0 & 3 \end{pmatrix} = \begin{pmatrix} 3 & 2 \\ 2 & 4 \end{pmatrix} \text{ and } \begin{pmatrix} 5 & 1 & 3 \\ -1 & 0 & x \end{pmatrix} + \begin{pmatrix} 2 & 1 & -3 \\ 3 & 2x & x \end{pmatrix} = \begin{pmatrix} 7 & 2 & 0 \\ 2 & 2x & 2x \end{pmatrix}.$$

In Example 6.1 above, the matrices A and B, A and C, B and C cannot be added.

We naturally denote $A + A$ by $2A$, so we define rA for any number r to be the matrix obtained from A by multiplying all the elements of A by r.

DON'T FORGET

Matrix addition looks like ordinary algebraic addition.
- $A + B = B + A$,
- $(A + B) + C$
  $= A + (B + C)$.

Example 6.4

$$2\begin{pmatrix} 1 & 3 \\ 2 & 1 \end{pmatrix} - 3\begin{pmatrix} 2 & -1 \\ 0 & 3 \end{pmatrix} = \begin{pmatrix} 2 & 6 \\ 4 & 2 \end{pmatrix} + \begin{pmatrix} -6 & 3 \\ 0 & -9 \end{pmatrix} = \begin{pmatrix} -4 & 9 \\ 4 & -7 \end{pmatrix}$$

MATRIX MULTIPLICATION

A matrix A can only be multiplied by another matrix B if the number of columns in A is the same as the number of rows in B. Multiplying a row matrix of order $1 \times n$ with a column matrix of order $n \times 1$, the product is given by:

$$n = 2: (a \quad b)\begin{pmatrix} c \\ d \end{pmatrix} = ac + bd; \quad n = 3: (a \quad b \quad c)\begin{pmatrix} d \\ e \\ f \end{pmatrix} = ad + be + cf; \text{ and so on.}$$

contd

MATRIX MULTIPLICATION   contd

In each case, the result is a number which is thought of as a 1×1 matrix.

In general, to define the product AB of two matrices, in the order written, the number of columns in A must equal the number of rows in B. Treat each row of A as a row matrix, and each column of B as a column matrix, and proceed as follows:

- Multiply row 1 of A by column 1 of B and put result in row 1 column 1.

- Multiply row 1 of A by column 2 of B and put result in row 1 column 2.

- When the columns of B have all been used, move to row 2 of A.

- Multiply row 2 of A by column 1 of B and put result in row 2 column 1.

- Continue until every row of A has been multiplied by every column of B.

The diagram below shows how to multiply a 2×3 matrix on the left by a 3×2 matrix on the right to produce a 2×2 matrix:

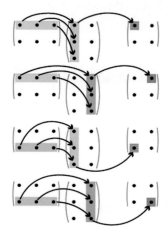

Example 6.5

Let $A = \begin{pmatrix} 1 & 2 \\ 0 & 1 \end{pmatrix}$, $B = \begin{pmatrix} 2 & 1 \\ 4 & 3 \end{pmatrix}$

Then $AB = \begin{pmatrix} 1 \times 2 + 2 \times 4 & 1 \times 1 + 2 \times 3 \\ 0 \times 2 + 1 \times 4 & 0 \times 1 + 1 \times 3 \end{pmatrix} = \begin{pmatrix} 10 & 7 \\ 4 & 3 \end{pmatrix}$

$BA = \begin{pmatrix} 2 \times 1 + 1 \times 0 & 2 \times 2 + 1 \times 1 \\ 4 \times 1 + 3 \times 0 & 4 \times 2 + 3 \times 1 \end{pmatrix} = \begin{pmatrix} 2 & 5 \\ 4 & 11 \end{pmatrix}$

Note that $AB \neq BA$.

Example 6.6

$\begin{pmatrix} 2 & 0 & 1 \\ -2 & 3 & 1 \\ 1 & -1 & 0 \end{pmatrix}\begin{pmatrix} 2 \\ 1 \\ 2 \end{pmatrix} = \begin{pmatrix} 2 \times 2 + 0 \times 1 + 1 \times 2 \\ (-2) \times 2 + 3 \times 1 + 1 \times 2 \\ 1 \times 2 + (-1) \times 1 + 0 \times 2 \end{pmatrix} = \begin{pmatrix} 6 \\ 1 \\ 1 \end{pmatrix}$

Note: when calculating AB, all answers to multiplying by row 1 of A go in row 1 in the resultant matrix, all for row 2 of A go in row 2, and so on. Similarly, all answers to multiplying by column 1 of B go in column 1 of the resultant matrix, and so on.

> **DON'T FORGET**
>
> Order matters when multiplying matrices.

contd

MATRICES

MATRIX MULTIPLICATION   contd

Although $AB \neq BA$ in general, for some matrices $AB = BA$.

Example 6.7

$$\begin{pmatrix} 1 & -1 \\ 2 & 0 \end{pmatrix}\begin{pmatrix} 2 & 1 \\ -2 & 3 \end{pmatrix} = \begin{pmatrix} 4 & -2 \\ 4 & 2 \end{pmatrix} = \begin{pmatrix} 2 & 1 \\ -2 & 3 \end{pmatrix}\begin{pmatrix} 1 & -1 \\ 2 & 0 \end{pmatrix}$$

Matrix algebra looks very much like ordinary algebra.

In general, $AB \neq BA$, but

$$(AB)C = A(BC)$$

and $A(B + C) = AB + AC$

and $AA = A^2$.

THE TRANSPOSE OF A MATRIX

If we interchange the rows and columns of a matrix A, the resulting matrix is denoted by A', or A^T, and called the **transpose** of A.

Example 6.8

If $A = \begin{pmatrix} 1 & 0 & 2 \\ 2 & 1 & 3 \end{pmatrix}$, $B = \begin{pmatrix} 2 & 0 \\ -1 & 5 \end{pmatrix}$, $C = \begin{pmatrix} 1 & 3 & 0 \end{pmatrix}$,

then $A' = \begin{pmatrix} 1 & 2 \\ 0 & 1 \\ 2 & 3 \end{pmatrix}$, $B' = \begin{pmatrix} 2 & -1 \\ 0 & 5 \end{pmatrix}$, $C' = \begin{pmatrix} 1 \\ 3 \\ 0 \end{pmatrix}$.

Points to note:

$$(A')' = A$$

$$(A + B)' = A' + B'$$

$$(AB)' = B'A'. \text{ Note the order on the right-hand side.}$$

Check this for the matrices A, B in Example 6.5.

If $A = A'$ for a matrix A, we say that A is symmetric.

Example 6.9

$\begin{pmatrix} 2 & 3 & 5 \\ 3 & 0 & 6 \\ 5 & 6 & -1 \end{pmatrix}$ is symmetric.

We can reflect in the leading diagonal: $\begin{pmatrix} 2 & & \\ & 0 & \\ & & -1 \end{pmatrix}$

LET'S THINK ABOUT THIS

If a matrix is symmetric, why must it be a square matrix?

Answer: If A has order $m \times n$, then A' has order $n \times m$, so they can only be equal when $m = n$.

NOW TRY THIS

(1) Given that A, B, C, D are square matrices, where

$$A = \begin{pmatrix} 2 & -1 \\ 3 & 5 \end{pmatrix}, \quad B = \begin{pmatrix} 4 & 6 \\ 0 & -3 \end{pmatrix}, \quad C = \begin{pmatrix} x & 2 \\ 0 & y \end{pmatrix}, \quad D = \begin{pmatrix} 2 & 7 \\ 12 & -1 \end{pmatrix}$$

(a) Find AB.   1

(b) Express $4C + D$ as a single matrix.   2

(c) Given that $AB = 4C + D$, find the values of x and y.   2

(2) Let $A = \begin{pmatrix} 1 & 1 & 1 \\ 1 & 2 & 3 \\ 1 & -1 & -1 \end{pmatrix}$ and $B = \begin{pmatrix} 1 & 0 & 1 \\ 4 & -2 & -2 \\ -3 & 2 & 1 \end{pmatrix}$.

Show that $AB = \begin{pmatrix} k & 0 & 0 \\ 0 & k & 0 \\ 0 & 0 & k \end{pmatrix}$ for some constant k.

Hence obtain the matrix A^2B.   3

(3) Given the matrix $A = \begin{pmatrix} t+4 & 3t \\ 3 & 5 \end{pmatrix}$

(a) Find A^{-1} in terms of t when A is non-singular.   3

(b) Write down the value of t such that A is singular.   1

(c) Given that the transpose of A is $\begin{pmatrix} 6 & 3 \\ 6 & 5 \end{pmatrix}$, find t.   1

SPECIAL MATRICES

IDENTITY MATRIX

For 2×2 matrices, the matrix $I = \begin{pmatrix} 1 & 0 \\ 0 & 1 \end{pmatrix}$ satisfies $AI = IA = A$ for **all** 2×2 matrices A.

I is called the **identity** (or **unit**) matrix.

> **DON'T FORGET**
>
> The identity matrix plays the part of the number 1 in multiplication.

For 3×3 matrices, the matrix $I = \begin{pmatrix} 1 & 0 & 0 \\ 0 & 1 & 0 \\ 0 & 0 & 1 \end{pmatrix}$ satisfies $AI = IA = A$ for **all** 3×3 matrices A.

Note that we use the letter I for both the 2×2 case and the 3×3 case. The context makes it clear which one we mean.

INVERSES OF SQUARE MATRICES

For a given square matrix A, if we can find a matrix B such that $AB = I$, B is called the inverse of A, and is denoted by A^{-1}. When A^{-1} exists, we have $AA^{-1} = I = A^{-1}A$.

> **DON'T FORGET**
>
> $AA^{-1} = I = A^{-1}A$

Inverses of 2 × 2 matrices

The expression $D = ad - bc$ is called the **determinant** of the matrix $\begin{pmatrix} a & b \\ c & d \end{pmatrix}$.

The determinant of a matrix A is often denoted by **det A** or by $\begin{vmatrix} a & b \\ c & d \end{vmatrix}$.

> **DON'T FORGET**
>
> If $A = \begin{pmatrix} a & b \\ c & d \end{pmatrix}$
> $\det A = ad - bc$
> $A^{-1} = \dfrac{1}{\det A}\begin{pmatrix} d & -b \\ -c & a \end{pmatrix}$

The inverse of A is given by $A^{-1} = \dfrac{1}{D}\begin{pmatrix} d & -b \\ -c & a \end{pmatrix}$.

If $D = 0$, then the matrix A doesn't have an inverse; A is said to be singular.

Example 6.10

Let $A = \begin{pmatrix} 2 & 1 \\ -1 & 3 \end{pmatrix}$ and $B = \begin{pmatrix} 2 & 1 \\ 4 & 2 \end{pmatrix}$.

Then $\det A = 2 \times 3 - (-1) \times 1 = 7$, and $\det B = 2 \times 2 - 4 \times 1 = 0$.

> **DON'T FORGET**
>
> $\det A = 0 \Leftrightarrow A$ is singular $\Leftrightarrow A^{-1}$ doesn't exist

So, A is non-singular, and $A^{-1} = \dfrac{1}{7}\begin{pmatrix} 3 & -1 \\ 1 & 2 \end{pmatrix} = \begin{pmatrix} \frac{3}{7} & -\frac{1}{7} \\ \frac{1}{7} & \frac{2}{7} \end{pmatrix}$.

B is singular; it doesn't have an inverse.

Inverses of 3 × 3 matrices

The determinant of a 3×3 matrix determines when the matrix has an inverse, just as in the 2×2 case.

If $A = \begin{pmatrix} a_1 & b_1 & c_1 \\ a_2 & b_2 & c_2 \\ a_3 & b_3 & c_3 \end{pmatrix}$ then $\det A = a_1\begin{vmatrix} b_2 & c_2 \\ b_3 & c_3 \end{vmatrix} - b_1\begin{vmatrix} a_2 & c_2 \\ a_3 & c_3 \end{vmatrix} + c_1\begin{vmatrix} a_2 & b_2 \\ a_3 & b_3 \end{vmatrix}$.

So, $\det A = a_1(b_2c_3 - b_3c_2) - b_1(a_2c_3 - a_3c_2) + c_1(a_2b_3 - a_3b_2)$.

Note the negative sign before the term in b_1.

contd

INVERSES OF SQUARE MATRICES   contd

Example 6.11

$$\begin{vmatrix} 2 & 1 & 2 \\ 1 & -1 & 0 \\ 3 & 2 & 2 \end{vmatrix} = 2\begin{vmatrix} -1 & 0 \\ 2 & 2 \end{vmatrix} - 1\begin{vmatrix} 1 & 0 \\ 3 & 2 \end{vmatrix} + 2\begin{vmatrix} 1 & -1 \\ 3 & 2 \end{vmatrix} = 2(-2-0) - (2-0) + 2(2-(-3)) = 4.$$

DON'T FORGET

$\det(AB) = \det A \det B$

Just as for 2×2 matrices (in fact, for all square matrices), a 3×3 matrix has an inverse if, and only if, its determinant is non-zero.

To calculate the inverse of a 3×3 matrix (when it has an inverse), we use elementary row operations as follows.

Start with the augmented matrix $A\vdots I$ and reduce it to the form $I\vdots B$ by elementary row operations. Then $B = A^{-1}$.

In $A\vdots I$, start by reducing A to A_1, which is in upper triangular form (I changes to I_1 when you do this), then reduce A_1 to lower triangular form A_2 (I_1 changes to I_2). Now reduce A_2 to I, changing I_2 to B, the inverse of A.

There are lots of ways in which elementary row operations can be used to get to the final form, so the steps in Example 6.12 below are not the only ones. There are also other valid methods for finding the inverse.

Example 6.12

Calculate the inverse of $\begin{pmatrix} 2 & 1 & 2 \\ 1 & -1 & 0 \\ 3 & 2 & 2 \end{pmatrix}$.

(We worked horizontally here to save space, but your working should progress **down** the page.)

$$\left[\begin{array}{ccc:ccc} 2 & 1 & 2 & 1 & 0 & 0 \\ 1 & -1 & 0 & 0 & 1 & 0 \\ 3 & 2 & 2 & 0 & 0 & 1 \end{array}\right] \rightarrow \left[\begin{array}{ccc:ccc} 2 & 1 & 2 & 1 & 0 & 0 \\ 0 & -3 & -2 & -1 & 2 & 0 \\ 0 & 1 & -2 & -3 & 0 & 2 \end{array}\right] \begin{array}{l} \\ 2R_2 - R_1 \\ 2R_3 - 3R_1 \end{array}$$

$$\left[\begin{array}{ccc:ccc} 2 & 1 & 2 & 1 & 0 & 0 \\ 0 & -3 & -2 & -1 & 2 & 0 \\ 0 & 0 & -8 & -10 & 2 & 6 \end{array}\right] \begin{array}{l} \\ \\ 3R_3 + R_2 \end{array} \rightarrow \left[\begin{array}{ccc:ccc} 8 & 4 & 0 & -6 & 2 & 6 \\ 0 & -12 & 0 & 6 & 6 & -6 \\ 0 & 0 & -8 & -10 & 2 & 6 \end{array}\right] \begin{array}{l} 4R_1 + R_3 \\ 4R_2 - R_3 \\ \end{array}$$

$$\left[\begin{array}{ccc:ccc} 24 & 0 & 0 & -12 & 12 & 12 \\ 0 & -12 & 0 & 6 & 6 & -6 \\ 0 & 0 & -8 & -10 & 2 & 6 \end{array}\right] \begin{array}{l} 3R_1 + R_2 \\ \\ \end{array}$$

Now we just divide row 1 by 24, row 2 by −12, and row 3 by −8 to give:

$$\left[\begin{array}{ccc:ccc} 1 & 0 & 0 & -\frac{1}{2} & \frac{1}{2} & \frac{1}{2} \\ 0 & 1 & 0 & -\frac{1}{2} & -\frac{1}{2} & \frac{1}{2} \\ 0 & 0 & 1 & \frac{5}{4} & -\frac{1}{4} & -\frac{3}{4} \end{array}\right]$$

So, the inverse of $\begin{pmatrix} 2 & 1 & 2 \\ 1 & -1 & 0 \\ 3 & 2 & 2 \end{pmatrix}$ is $\begin{pmatrix} -\frac{1}{2} & \frac{1}{2} & \frac{1}{2} \\ -\frac{1}{2} & -\frac{1}{2} & \frac{1}{2} \\ \frac{5}{4} & -\frac{1}{4} & -\frac{3}{4} \end{pmatrix}$.

contd

MATRICES

If you reduce $A \vdots I$ to $E \vdots F$, where

$$E = \begin{matrix} p & 0 & 0 \\ 0 & q & 0 \\ 0 & 0 & r \end{matrix}, \text{ and } p, q, r \text{ are integers,}$$

you can avoid fractions until the last step.

Because it requires so much arithmetic to obtain the inverse of a 3×3 matrix, many examination questions do not require you to use elementary row operations. The next example is typical.

Example 6.13

Let $A = \begin{pmatrix} 1 & 2 & 1 \\ 2 & -1 & 1 \\ 1 & 0 & 2 \end{pmatrix}$ and $B = \begin{pmatrix} 2 & 4 & -3 \\ 3 & x & -1 \\ x & -2 & 5 \end{pmatrix}$. Calculate AB and hence, or otherwise, obtain A^{-1}.

$$AB = \begin{pmatrix} 8+x & 2+2x & 0 \\ x+1 & 6-x & 0 \\ 2+2x & 0 & 7 \end{pmatrix}.$$

If we choose $x = -1$, we get $\quad AB = \begin{pmatrix} 7 & 0 & 0 \\ 0 & 7 & 0 \\ 0 & 0 & 7 \end{pmatrix} = 7I \quad$ and $\quad B = \begin{pmatrix} 2 & 4 & -3 \\ 3 & -1 & -1 \\ -1 & -2 & 5 \end{pmatrix}$.

Hence $A^{-1} = \frac{1}{7}B = \begin{pmatrix} \frac{2}{7} & \frac{4}{7} & -\frac{3}{7} \\ \frac{3}{7} & -\frac{1}{7} & -\frac{1}{7} \\ -\frac{1}{7} & -\frac{2}{7} & \frac{5}{7} \end{pmatrix}$.

In questions like this, there are usually only 1 or 2 marks available for obtaining A^{-1}, so although you **could** use elementary row operations to obtain it, you should be on the lookout for the quicker method.

Example 6.14

Given two matrices A and B of the same order (any order), and given A^{-1}, B^{-1}, how do we get the inverses of AB and BA?

We have $(AB)(B^{-1}A^{-1}) = A(BB^{-1})A^{-1} = AIA^{-1} = AA^{-1} = I$, because brackets can go anywhere in a product (but don't change the order of the product).

This shows that the inverse of AB is $B^{-1}A^{-1}$ (note the order).

In the same way, the inverse of BA is $A^{-1}B^{-1}$.

DON'T FORGET

$(AB)^{-1} = B^{-1}A^{-1}$

 LET'S THINK ABOUT THIS

DON'T FORGET

You cannot divide matrices.

Why can't we define $\dfrac{A}{B}$ for matrices A and B, even when B is non-singular?

Answer: We could (but we don't) define $\dfrac{1}{B}$ to be B^{-1}, but then we wouldn't know whether $\dfrac{A}{B}$ meant AB^{-1} or $B^{-1}A$.

NOW TRY THIS

(1) Let the matrix $A = \begin{pmatrix} 1 & x \\ x & 4 \end{pmatrix}$.

    (a) Obtain the value(s) of x for which A is singular.   **2**

    (b) When $x = 2$, show that $A^2 = pA$ for some constant p.

        Determine the value of q such that $A^4 = qA$.   **3**

(2) Calculate the inverse of the matrix $\begin{pmatrix} 1 & 2 \\ -x & 3 \end{pmatrix}$.

    For what value of x is this matrix singular?   **4**

(3) Matrices A and B are defined by

$$A = \begin{pmatrix} 1 & 0 & -1 \\ 0 & 1 & -1 \\ 0 & 1 & 2 \end{pmatrix}, \quad B = \begin{pmatrix} x+2 & x-2 & x+3 \\ -4 & 4 & 2 \\ 2 & -2 & 3 \end{pmatrix}.$$

    (a) Find the product AB.   **2**

    (b) Obtain the determinants of A and of AB.   **2**

        Hence, or otherwise, obtain an expression for det B.   **1**

(4) Given the matrix $A = \begin{pmatrix} \lambda & 2 \\ \lambda+3 & 4 \end{pmatrix}$,

    (a) Obtain A^{-1} when A is non-singular.   **3**

    (b) For what value of λ is A singular?   **1**

    (c) Given that $A' = \begin{pmatrix} -2 & 1 \\ 2 & 4 \end{pmatrix}$, obtain the value of λ.   **1**

(5) Determine k such that the matrix $\begin{pmatrix} 1 & 1 & 0 \\ 0 & k-2 & -1 \\ 1 & 2 & k \end{pmatrix}$ does not have an inverse.   **4**

(6) (a) Given the matrix $A = \begin{pmatrix} 0 & 4 & 2 \\ 1 & 0 & 1 \\ -1 & -2 & -3 \end{pmatrix}$, show that $A^2 + A = kI$ for some constant k,

       where I is the 3×3 unit matrix.   **4**

    (b) Obtain the values of p and q for which $A^{-1} = pA + qI$.   **2**

LINEAR EQUATIONS IN MATRIX FORM

The system of linear equations $\quad a_1x + b_1y = c_1 \qquad a_2x + b_2y = c_2$

is equivalent to the matrix equation: $\quad \begin{pmatrix} a_1 & b_1 \\ a_2 & b_2 \end{pmatrix}\begin{pmatrix} x \\ y \end{pmatrix} = \begin{pmatrix} c_1 \\ c_2 \end{pmatrix}$,

that is, $A\begin{pmatrix} x \\ y \end{pmatrix} = \begin{pmatrix} c_1 \\ c_2 \end{pmatrix}$ where $A = \begin{pmatrix} a_1 & b_1 \\ a_2 & b_2 \end{pmatrix}$.

So, when A is non-singular, multiplying on the left by A^{-1} gives $A^{-1}\left(A\begin{pmatrix} x \\ y \end{pmatrix}\right) = A^{-1}\begin{pmatrix} c_1 \\ c_2 \end{pmatrix}$.

But $A^{-1}\left(A\begin{pmatrix} x \\ y \end{pmatrix}\right) = (A^{-1}A)\begin{pmatrix} x \\ y \end{pmatrix}$.

So, $A^{-1}\left(A\begin{pmatrix} x \\ y \end{pmatrix}\right) = (A^{-1}A)\begin{pmatrix} x \\ y \end{pmatrix} = I\begin{pmatrix} x \\ y \end{pmatrix} = \begin{pmatrix} x \\ y \end{pmatrix}$,

> **DON'T FORGET**
>
> Brackets can go anywhere in a product.

hence $\begin{pmatrix} x \\ y \end{pmatrix} = A^{-1}\begin{pmatrix} c_1 \\ c_2 \end{pmatrix}$, from which we get x and y.

So, if we can find A^{-1}, the system of linear equations can be solved.

This works for the 3×3 case also.

Example 6.15

Given that the inverse of $A = \begin{pmatrix} 1 & 2 & 1 \\ 2 & -1 & 1 \\ 1 & 0 & 2 \end{pmatrix}$ is $\begin{pmatrix} \frac{2}{7} & \frac{4}{7} & -\frac{3}{7} \\ \frac{3}{7} & -\frac{1}{7} & -\frac{1}{7} \\ -\frac{1}{7} & -\frac{2}{7} & \frac{5}{7} \end{pmatrix}$,

solve the system of equations:

$x + 2y + z = 14, \quad 2x - y + z = 7, \quad x + 2z = -21.$

> **DON'T FORGET**
>
> Unique solution
> $\Leftrightarrow \det A \neq 0$

$$\begin{pmatrix} x \\ y \\ z \end{pmatrix} = A^{-1}\begin{pmatrix} 14 \\ 7 \\ -21 \end{pmatrix} = \begin{pmatrix} \frac{2}{7} & \frac{4}{7} & -\frac{3}{7} \\ \frac{3}{7} & -\frac{1}{7} & -\frac{1}{7} \\ -\frac{1}{7} & -\frac{2}{7} & \frac{5}{7} \end{pmatrix}\begin{pmatrix} 14 \\ 7 \\ -21 \end{pmatrix} = \begin{pmatrix} 4+4+9 \\ 6-1+3 \\ -2-2-15 \end{pmatrix} = \begin{pmatrix} 17 \\ 8 \\ -19 \end{pmatrix}$$

Hence $x = 17, \quad y = 8, \quad z = -19.$

A system of linear equations has a unique solution if, and only if, the matrix of coefficients of the unknowns is non-singular.

GEOMETRIC TRANSFORMATIONS OF THE PLANE **Unit 3**

Transformations of the plane, such as rotations and reflections, can be represented by matrices.
If the transformation maps the point (1,0) to (a,c) and (0,1) to (b,d), it is represented by the matrix
$\begin{pmatrix} a & b \\ c & d \end{pmatrix}$. The point (1,0) is represented by the matrix $\begin{pmatrix} 1 \\ 0 \end{pmatrix}$ and the point (0,1) by the matrix $\begin{pmatrix} 0 \\ 1 \end{pmatrix}$.

$$\begin{pmatrix} a & b \\ c & d \end{pmatrix}\begin{pmatrix} x \\ y \end{pmatrix} = \begin{pmatrix} ax+by \\ cx+dy \end{pmatrix}$$

$$\begin{pmatrix} 1 & 0 \\ 0 & 1 \end{pmatrix} \text{ goes to } \begin{pmatrix} a & b \\ c & d \end{pmatrix}$$
$$\uparrow \quad \uparrow$$
$$(1,0)\ (0,1)$$

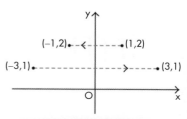

Hence the point $(x,y) \rightarrow (ax+by,\ cx+dy)$ under the transformation.

$$\begin{pmatrix} -1 & 0 \\ 0 & 1 \end{pmatrix}\begin{pmatrix} x \\ y \end{pmatrix} = \begin{pmatrix} -x \\ y \end{pmatrix}$$
So $(x,y) \rightarrow (-x,y)$, i.e. reflection in the y-axis.

If the matrices A, B represent transformations T_1, T_2 of the plane, then BA represents the transformation given by T_1 followed by T_2. Note the order of the matrices!

DON'T FORGET

BA means A first.

Anticlockwise rotation about origin through an angle θ	$\begin{pmatrix} \cos\theta & -\sin\theta \\ \sin\theta & \cos\theta \end{pmatrix}$
Reflection in the x-axis	$\begin{pmatrix} 1 & 0 \\ 0 & -1 \end{pmatrix}$
Reflection in the y-axis	$\begin{pmatrix} -1 & 0 \\ 0 & 1 \end{pmatrix}$
Reflection in the line $y = x$	$\begin{pmatrix} 0 & 1 \\ 1 & 0 \end{pmatrix}$
Reflection in the line $y = -x$	$\begin{pmatrix} 0 & -1 \\ -1 & 0 \end{pmatrix}$
Reflection in the origin	$\begin{pmatrix} -1 & 0 \\ 0 & -1 \end{pmatrix}$
Stretching ($k > 1$) or reduction ($k < 1$)	$\begin{pmatrix} k & 0 \\ 0 & k \end{pmatrix}$

Example 6.16

Write down the matrices A, B representing a reflection in the y-axis and an anticlockwise rotation about the origin through 45°. Hence obtain the matrices representing the result of applying:

(1) the reflection followed by the rotation

(2) the rotation followed by the reflection.

$$A = \begin{pmatrix} -1 & 0 \\ 0 & 1 \end{pmatrix}, \quad B = \begin{pmatrix} \frac{1}{\sqrt{2}} & -\frac{1}{\sqrt{2}} \\ \frac{1}{\sqrt{2}} & \frac{1}{\sqrt{2}} \end{pmatrix}$$

(1) $\begin{pmatrix} \frac{1}{\sqrt{2}} & -\frac{1}{\sqrt{2}} \\ \frac{1}{\sqrt{2}} & \frac{1}{\sqrt{2}} \end{pmatrix}\begin{pmatrix} -1 & 0 \\ 0 & 1 \end{pmatrix} = \begin{pmatrix} -\frac{1}{\sqrt{2}} & -\frac{1}{\sqrt{2}} \\ -\frac{1}{\sqrt{2}} & \frac{1}{\sqrt{2}} \end{pmatrix}$

(2) $\begin{pmatrix} -1 & 0 \\ 0 & 1 \end{pmatrix}\begin{pmatrix} \frac{1}{\sqrt{2}} & -\frac{1}{\sqrt{2}} \\ \frac{1}{\sqrt{2}} & \frac{1}{\sqrt{2}} \end{pmatrix} = \begin{pmatrix} -\frac{1}{\sqrt{2}} & \frac{1}{\sqrt{2}} \\ \frac{1}{\sqrt{2}} & \frac{1}{\sqrt{2}} \end{pmatrix}.$

Note that the resulting transformations are different.

NOW TRY THIS

(1) (a) Write down the 2×2 matrix M_1 associated with an anticlockwise rotation of $\frac{\pi}{2}$ radians about the origin. **2**

    (b) Write down the matrix M_2 associated with reflection in the x-axis. **1**

    (c) Evaluate $M_2 M_1$ and describe geometrically the effect of the transformation represented by $M_2 M_1$. **2**

(2) Obtain the 2×2 matrix M associated with an enlargement, scale factor 2, followed by a clockwise rotation of 60° about the origin. **4**

http://tutorial.math.lamar.edu/Classes/Alg/ComplexNumbers.aspx
http://mathworld.wolfram.com/ComplexNumber.html

INTRODUCING COMPLEX NUMBERS

The equation $x^2 + 1 = 0$ has no solutions in $\mathbb{R}$, the set of real numbers. The solution is $\sqrt{-1}$ and is denoted by i.

A *complex number* is of the form $z = a + bi$, where a and b are real numbers. When $b = 0$, we have the real number a.

The set of all complex numbers is denoted by $\mathbb{C}$. $\mathbb{R}$ is a subset of $\mathbb{C}$.

Complex numbers obey the rules of algebra.

DON'T FORGET

$i^2 = -1,\ i^3 = -i,\ i^4 = 1$

Example 7.1

If $z = 2 + i$, obtain z^2 and z^3.

$$z^2 = (2+i)^2 = 4 + 4i + i^2 = 4 + 4i - 1 = 3 + 4i.$$

$$z^3 = z^2 z = (3 + 4i)(2 + i) = 6 + 3i + 8i + 4i^2 = 6 + 11i - 4 = 2 + 11i.$$

DON'T FORGET

If $z = a + bi$, then $\bar{z} = a - bi$

If $z = a + bi$, the complex number $a - bi$ is called the **conjugate** of z and is denoted by $\bar{z}$.

Multiplying z by its conjugate $\bar{z}$ gives:

$$z\bar{z} = (a + bi)(a - bi)$$
$$= a^2 - abi + abi - b^2 i = a^2 + b^2.$$

We use this property of the complex conjugate to carry out division.

Example 7.2

If $z = 2 + i$, obtain $\dfrac{1}{z}$ in the form $a + bi$.

$$\frac{1}{z} = \frac{1}{2+i} \qquad = \frac{1}{2+i} \times \frac{2-i}{2-i}$$

$$= \frac{2-i}{4-i^2} = \frac{2-i}{5} = \frac{2}{5} - \frac{i}{5}$$

Two complex numbers $z = a + bi$ and $w = c + di$ are equal if, and only if, $a = c$ **and** $b = d$.

When working with complex numbers, you will sometimes be required to equate real and imaginary parts.

Example 7.3

Obtain the square roots of $3 - 4i$, i.e. solve the equation $z^2 = 3 - 4i$.

Let $z = x + yi$, so that $z^2 = x^2 + 2xyi + y^2 i^2 = x^2 - y^2 + 2xyi$.

So, $x^2 - y^2 + 2xyi = 3 - 4i \Rightarrow x^2 - y^2 = 3$ and $2xy = -4$.

Hence $y = -\dfrac{2}{x}$ and $x^2 - \dfrac{4}{x^2} = 3$.

Multiplying through by x^2 gives: $x^4 - 3x^2 - 4 = 0$,

and factorising gives: $(x^2 - 4)(x^2 + 1) = 0$, which gives:

$x = \pm 2$ because the second factor is always positive (remember that x is real).

$x = 2$ gives $y = -1$ and $x = -2$ gives $y = 1$.

Hence the square roots of $3 - 4i$ are $2 - i$ and $-2 + i$.

THE ARGAND DIAGRAM

The Argand diagram is a geometric way of representing complex numbers.
The diagram shows a general complex number $z = a + bi$.

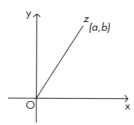

The distance of z from O in the diagram is called the **modulus** of z, denoted by $|z|$.

If $z = a + bi$, then $|z| = +\sqrt{a^2 + b^2}$.

Note that $z\bar{z} = |z|^2$.

The angle (in radians) that the line joining O to z makes with the positive x-axis is called the **argument** of z, denoted by $\arg z$.

If $z = a + bi$, then $\arg z = \tan^{-1}\dfrac{b}{a}$ $(a \neq 0)$.

Note that, in the 3rd and 4th quadrants, $\arg z$ is given as a negative angle, where the imaginary part of z is negative.

If $z = bi$, then $\arg z = \dfrac{\pi}{2}$ when $b > 0$ and $\arg z = -\dfrac{\pi}{2}$ when $b < 0$.

Example 7.4

If $z = -1 + \sqrt{3}i$, show z and $\bar{z}$ on an Argand diagram. Calculate $|z|$ and $\arg z$.

$|z| = +\sqrt{(-1)^2 + (\sqrt{3})^2} = \sqrt{4} = 2$

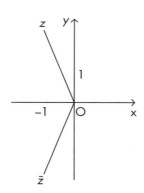

From the Argand diagram, we see that $\arg z$ will be an angle between $\dfrac{\pi}{2}$ and π:

$\arg z = \pi - \tan^{-1}(\sqrt{3}) = \pi - \dfrac{\pi}{3} = \dfrac{2\pi}{3}$.

> **DON'T FORGET**
>
> Usually, $-\pi < \arg z \leqslant \pi$

> **DON'T FORGET**
>
> $z = a + bi$ gives
> $|z| = +\sqrt{a^2 + b^2}$

> **DON'T FORGET**
>
> $z = a + bi$ gives
> $\arg z = \tan^{-1}\dfrac{b}{a}$ $(a \neq 0)$

NOW TRY THIS

(1) Express the inverse of $z = 2 - i$ in the form $a + bi$.

Plot z and z^2 on an Argand diagram.   **3**

(2) Solve the equation $z + |z|^2 = 7 - i$, giving your answers in the form $z = a + bi$.   **4**

(3) Given $z = 2 + ai$, where $a > 0$, obtain the value of a given that z^3 is real.

Evaluate $|z|$ and $\arg z$.   **6**

POLAR FORM and de MOIVRE'S THEOREM

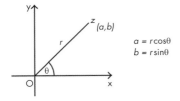

We can also use polar coordinates to represent complex numbers.

In this form, $z = r(\cos\theta + i\sin\theta)$ where $r = |z|$ and $\theta = \arg z$.

Products are easy in this form:

If $z_1 = r_1(\cos\theta_1 + i\sin\theta_1)$ and $z_2 = r_2(\cos\theta_2 + i\sin\theta_2)$, then:

$z_1 z_2 = r_1 r_2(\cos(\theta_1 + \theta_2) + i\sin(\theta_1 + \theta_2))$.

This shows that $|z_1 z_2| = |z_1||z_2|$ and $\arg(z_1 z_2) = \arg(z_1) + \arg(z_2)$ (with a possible adjustment if this sum lies outside the range $-\pi < \theta \leq \pi$).

We note also that, when the argument is negative, the complex number z can be written in a slightly different way.

Using trig properties of negative angles:

$z = r(\cos(-\theta) + i\sin(-\theta)) = r(\cos\theta - i\sin\theta)$.

If we take $z = \cos\theta + i\sin\theta$, then $z^n = \cos n\theta + i\sin n\theta$.
This can be written as: $(\cos\theta + i\sin\theta)^n = \cos n\theta + i\sin n\theta$.

This is **de Moivre's theorem**. It is valid for any real number n.

> **DON'T FORGET**
>
> If $z = r(\cos\theta + i\sin\theta)$, then $z^n = r^n(\cos n\theta + i\sin n\theta)$.

Example 7.5

Express $\sqrt{3}+i$ in polar form and hence express $(\sqrt{3}+i)^8$ in the form $a + bi$.

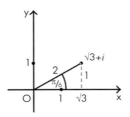

We have $\sqrt{3}+i = 2(\cos\dfrac{\pi}{6} + i\sin\dfrac{\pi}{6})$, so:

$(\sqrt{3}+i)^8 = 2^8(\cos\dfrac{8\pi}{6} + i\sin\dfrac{8\pi}{6})$.

$\dfrac{8\pi}{6} = \dfrac{4\pi}{3}$ and $\cos\dfrac{4\pi}{3} = -\dfrac{1}{2}$, $\sin\dfrac{4\pi}{3} = -\dfrac{\sqrt{3}}{2}$, so:

$(\sqrt{3}+i)^8 = 2^8(-\dfrac{1}{2} - \dfrac{\sqrt{3}}{2}i) = -2^7 - 2^7\sqrt{3}\,i$.

Example 7.6

Use de Moivre's theorem to prove that:

$$\cos 3\theta = 4\cos^3\theta - 3\cos\theta \qquad \text{and} \qquad \sin 3\theta = 3\sin\theta - 4\sin^3\theta.$$

Using de Moivre's theorem, we know $(\cos\theta + i\sin\theta)^3 = \cos 3\theta + i\sin 3\theta$.

Also, expanding the left-hand side by the binomial theorem:
$(\cos\theta + i\sin\theta)^3 = \cos^3\theta + 3i\cos^2\theta\sin\theta + 3i^2\cos\theta\sin^2\theta + i^3\sin^3\theta$.

Because $i^2 = -1$, $i^3 = -i$, this simplifies to:
$(\cos\theta + i\sin\theta)^3 = \cos^3\theta + 3i\cos^2\theta\sin\theta - 3\cos\theta\sin^2\theta - i\sin^3\theta$.

Hence:

$\cos 3\theta + i\sin 3\theta = \cos^3\theta - 3\cos\theta\sin^2\theta + 3i\cos^2\theta\sin\theta - i\sin^3\theta$.

Equating the real parts on both sides, and also the imaginary parts:

$$\cos 3\theta = \cos^3\theta - 3\cos\theta\sin^2\theta \qquad \text{and} \qquad \sin 3\theta = 3\cos^2\theta\sin\theta - \sin^3\theta.$$

Finally, using $\cos^2\theta + \sin^2\theta = 1$ gives the results:

$$\cos 3\theta = 4\cos^3\theta - 3\cos\theta \qquad \text{and} \qquad \sin 3\theta = 3\sin\theta - 4\sin^3\theta.$$

Of course, these results can be obtained without using complex numbers, but many such results are best proved in this way.

nTH ROOTS OF COMPLEX NUMBERS

If z is an nth root of a given complex number z_0, then $z^n = z_0$. By the Fundamental Theorem of Algebra (see next page), this equation has n roots, so every complex number has n nth roots (remember, this includes real numbers as well).

Example 7.7

Use de Moivre's theorem to obtain the cube roots of -1.

We must solve the equation $z^3 = -1$, so begin by setting $z = r(\cos\theta + i\sin\theta)$.

Then $z^3 = r^3(\cos 3\theta + i\sin 3\theta)$, so $r^3\cos 3\theta + i\,r^3\sin 3\theta = -1$.

The modulus of the left-hand side is r^3, and for the right-hand side it is 1. Hence $r = 1$.

So, $\cos 3\theta + i\sin 3\theta = -1$, and equating real and imaginary parts on both sides gives $\cos 3\theta = -1$ and $\sin 3\theta = 0$.

The values for which $\cos 3\theta = -1$ **and** $\sin 3\theta = 0$ **where also** $-\pi < \theta \leqslant \pi$ are:

$$\theta = \pi, \quad \theta = \frac{\pi}{3}, \quad \theta = -\frac{\pi}{3}.$$

So, $z = \cos\pi + i\sin\pi$, $\quad z = \cos\dfrac{\pi}{3} + i\sin\dfrac{\pi}{3}$, $\quad z = \cos\dfrac{\pi}{3} - i\sin\dfrac{\pi}{3}$,

giving $z = -1$, $\quad z = \dfrac{1}{2} + \dfrac{\sqrt{3}}{2}i$, $\quad z = \dfrac{1}{2} - \dfrac{\sqrt{3}}{2}i$. $\qquad$ Note that the complex roots are conjugates.

NOW TRY THIS

(1) Use de Moivre's theorem to express z^k in terms of θ, where $z = \cos\theta + i\sin\theta$, and k is a positive integer.

   (a) Hence show that $z^{-k} = \cos k\theta - i\sin k\theta$, and obtain expressions for $\cos k\theta$ and $\sin k\theta$ in terms of z. **3**

   (b) Show that $\cos 2\theta \sin^2\theta = -\dfrac{1}{8}\left(z^2 + \dfrac{1}{z^2}\right)\left(z - \dfrac{1}{z}\right)^2$. **2**

   (c) Hence show that $\cos 2\theta \sin^2\theta = a + b\cos 2\theta + c\cos 4\theta$ for suitable constants a, b, c. **3**

(2) Let $z = r(\cos\theta + i\sin\theta)$.

   (a) Use the binomial theorem to express z^4 in the form $u + iv$, where u and v are expressions involving $\sin\theta$ and $\cos\theta$. **3**

   (b) Use de Moivre's theorem to write down a second expression for z^4. **1**

   (c) Using the results of (a) and (b), show that: $\dfrac{\cos 4\theta}{\cos^2\theta} = p\cos^2\theta + q\sec^2\theta + r$,

   where $-\dfrac{\pi}{2} < \theta < \dfrac{\pi}{2}$, stating the values of p, q and r. **6**

(3) Given $z = r(\cos\theta + i\sin\theta)$, use de Moivre's theorem to express z^3 in polar form. **1**

   Hence obtain $\left(\cos\dfrac{2\pi}{3} + i\sin\dfrac{2\pi}{3}\right)^3$ in the form $a + ib$. **2**

   Hence, or otherwise, obtain the roots of the equation $z^3 = 8$ in Cartesian form. **4**

POLYNOMIAL EQUATIONS

DON'T FORGET

If z is a root, then so is $\bar{z}$.

Once we have $\mathbb{C}$, it turns out that every polynomial equation of degree n has n roots, allowing for repeated roots. This is the **Fundamental Theorem of Algebra**.

When the coefficients are real, the complex roots (if any) occur in **conjugate pairs**.

Thus, if $3 + 2i$ is a root, then so is $3 - 2i$.

Example 7.8

Verify that $3 + i$ is a solution of the equation $z^3 - z^2 - 20z + 50 = 0$. Hence find all the solutions.

$$(3 + i)^2 = 8 + 6i, \quad (3 + i)^3 = (8 + 6i)(3 + i) = 18 + 26i.$$

Substituting into $z^3 - z^2 - 20z + 50$ gives:

$$(3 + i)^3 - (3 + i)^2 - 20(3 + i) + 50 = 18 + 26i - (8 + 6i) - 60 - 20i + 50 = 0 + 0i = 0.$$

Because $3 + i$ is a root, so is $3 - i$, and so:

$$(z - (3 + i))(z - (3 - i)) = (z - 3 - i)(z - 3 + i) = z^2 - 6z + 10 \text{ is a factor}$$
of $z^3 - z^2 - 20z + 50$.

Writing $z^3 - z^2 - 20z + 50 = (z^2 - 6z + 10)(z + a)$ gives $a = 5$.

An alternative is to divide $z^3 - z^2 - 20z + 50$ by $z^2 - 6z + 10$ to give $z + 5$.

Therefore, the roots of $z^3 - z^2 - 20z + 50 = 0$ are $z = -5, \quad 3 + i \quad$ and $\quad 3 - i$.

NOW TRY THIS

(1) Given $z = 1 + 2i$, express $z^2(z+3)$ in the form $a + bi$.   **2**

Hence, or otherwise, verify that $1 + 2i$ is a root of the equation

$$z^3 + 3z^2 - 5z + 25 = 0.$$ **2**

Obtain the other roots of this equation.   **2**

(2) Given that $3 - i$ is a root of the equation $2z^3 - 11z^2 + 14z + 10 = 0$, obtain all the roots.   **4**

(3) Verify that i is a solution of $z^4 + 4z^3 + 3z^2 + 4z + 2 = 0$.

Hence find all the solutions.   **5**

(4) Given that $i - 1$ is a root of the equation $f(z) = z^4 - z^3 - 5z^2 - 8z - 2 = 0$, obtain all the roots.   **5**

GEOMETRIC FIGURES IN THE COMPLEX PLANE

The expression $|z - z_0|$ gives the distance in the Argand diagram between the complex numbers z and z_0.

It follows that the equation $|z - z_0| = r$, where z_0 is fixed, and $r > 0$ is a given real number, gives all those z at a fixed distance r from z_0. In other words, the circle, centre z_0 with radius r.

The set of all z for which $|z - z_0| < r$ gives the interior of the circle centre z_0 with radius r.

Example 7.9

Describe the locus (meaning the position of all points satisfying the given condition) given by the equation $|z - i| = |z + 1|$.

Method 1

Let $z = x + yi$.

$|z - i|^2 = |x + i(y - 1)|^2 = x^2 + (y - 1)^2$.

Similarly, $|z + 1|^2 = |(x + 1) + yi|^2 = (x + 1)^2 + y^2$.

So, $x^2 + (y - 1)^2 = (x + 1)^2 + y^2$.

Expanding and simplifying gives $y = -x$, a straight line through the origin.

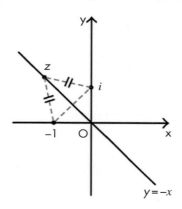

Method 2

The equation states that the distance of z from i equals its distance from -1.

Hence z lies on the perpendicular bisector of the line joining i and -1, which is the line $y = -x$.

NOW TRY THIS

(1) Identify the locus in the complex plane given by $|z + i| = 2$.   **3**

    Hence show the set of complex numbers z for which $|z + i| \leqslant 2$.   **2**

ARITHMETIC SEQUENCES

DON'T FORGET

The nth term of the arithmetic sequence with 1st term a and common difference d is:
$u_n = a + (n - 1)d$

In an arithmetic sequence, consecutive terms change by a constant amount: $u_1 = a$, $u_2 = a + d$, $u_3 = a + 2d$, ..., and generally: $u_n = a + (n - 1)d$.

Note that $u_{k+1} = u_k + d$, where d is the **common difference** (it may be positive or negative).

The associated series $S_n = a + (a + d) + (a + 2d) + ... + (a + [n - 1]d)$ is called an **arithmetic series**.

Example 8.1

The sequence 3, 5, 7, ... is an arithmetic sequence with $a = 3$ and $d = 2$.
We have: $u_n = 2n + 1$ ($n = 1, 2, 3, ...$).

DON'T FORGET

In an arithmetic sequence, add d to get the next term.

SUM OF AN ARITHMETIC SERIES

$$
\begin{aligned}
\text{Let } S_n &= a + (a + d) + (a + 2d) + ... + (a + [n - 1]d) \\
&= an + d(1 + 2 + 3 + ... + (n - 1)) \\
&= an + \frac{1}{2}n(n - 1)d \\
&= \frac{n}{2}\{2a + (n - 1)d\}
\end{aligned}
$$

At least one version of this formula must be remembered. An alternative form, which is sometimes useful, is:

$S_n = \frac{n}{2}(u_1 + u_n)$, where u_1 and u_n are the first and last terms in the sequence.

DON'T FORGET

$1 + 2 + 3 + ... + n = \frac{n(n+1)}{2}$

Example 8.2

$$\sum_{k=1}^{6} 4 = 6 \times 4 = 24$$

DON'T FORGET

$\sum_{k=1}^{n} c = nc$ for any constant c.

Example 8.3

Obtain a formula for $\sum_{k=1}^{n}(3k + 1)$.   Hence evaluate $301 + 304 + 307 + ... + 601$.

Method 1

This is an arithmetic series with $a = 4$ and $d = 3$. Writing out the first few terms, i.e. 4, 7, 10, ..., will make this clear.

We have $a = 4$, $d = 3$, so:

$$S_n = \frac{n}{2}\{8 + 3(n - 1)\} = \frac{1}{2}n(3n + 5).$$

Method 2

$$\sum_{k=1}^{n}(3k + 1) = 3\sum_{k=1}^{n}k + \sum_{k=1}^{n}1 = \frac{3n(n+1)}{2} + n$$

$$= \frac{n}{2}(3n + 3 + 2) = \frac{n}{2}(3n + 5)$$

Hence: $301 + 304 + 307 + ... + 601 = S_{200} - S_{99}$ since $k = 200$ gives the 601 term and $k = 99$ gives the term before 301.

So: $301 + 304 + 307 + ... + 601 = \dfrac{200 \times 605}{2} - \dfrac{99 \times 302}{2} = 45\,551$.

Alternatively, using $S_n = \frac{n}{2}(u_1 + u_n)$, we have $200 - 99 = 101$ terms and:

$$301 + 304 + 307 + ... + 601 = \frac{101}{2}(301 + 601) = 45\,551.$$

contd

SUM OF AN ARITHMETIC SERIES   contd

Example 8.4

(i)   Obtain a formula for the sum, S_n, of the first n terms of the sequence 9, 7, 5, …

(ii) Find the values of n for which $S_n = 21$.

(iii) Obtain the smallest value of n for which $S_n < -50$.

DON'T FORGET

Sum (arithmetic)
$= \frac{n}{2}\{2a + (n-1)d\}$

(i)   The sequence is arithmetic with $a = 9$ and $d = -2$, so:

$$S_n = \frac{n}{2}\{18 - 2(n-1)\} = \frac{n}{2}(20 - 2n) = 10n - n^2$$

(ii) $S_n = 21 \implies 10n - n^2 = 21$,   so $n^2 - 10n + 21 = 0$

$(n-3)(n-7) = 0$, so the sum is 21 when $n = 3$ and when $n = 7$.

(iii) We require the smallest value of n ($n \geqslant 1$) for which $10n - n^2 < -50 \implies n^2 - 10n > 50$.

Completing the square, $(n-5)^2 > 75$, so $n > 5 + \sqrt{75} \approx 13 \cdot 7$.

So, the smallest value of n is 14.

(An alternative to completing the square is to solve $x^2 - 10x - 50 = 0$, using the formula, and explain why (with a rough diagram, for example) the value of n is the smallest integer greater than the positive root of this quadratic.)

NOW TRY THIS

(1) The first term of an arithmetic sequence is 3 and the 10th term is –15.
Obtain the sum of the first 100 terms.   4

(2) Show that $\displaystyle\sum_{r=1}^{n}(6 + 2r) = 7n + n^2$.   2

Hence write down a formula for $\displaystyle\sum_{r=1}^{3k}(6 + 2r)$.   1

Show that $\displaystyle\sum_{r=k+1}^{3k}(6 + 2r) = 14k + 8k^2$.   2

(3) The sum $S(n)$ of the first n terms of a sequence $u_1, u_2, u_3, \ldots$ is given by:

$S(n) = 8n - n^2$, $n \geqslant 1$.

Calculate the values of u_1, u_2, u_3 and state what type of sequence it is.   3

Obtain a formula for u_n in terms of n, simplifying your answer.   2

GEOMETRIC SEQUENCES

 **DON'T FORGET**

In a geometric sequence, multiply by r to get the next term.

In a geometric sequence, consecutive terms change by a constant factor: a, ar, ar^2, ...

The factor r is called the **common ratio**, and can have any value (but when $r = 0$ or $r = 1$ the sequence becomes trivial).

Note that $u_k = ar^{k-1}$ and $u_{k+1} = ru_k$.

The partial sums of geometric sequences are called **geometric series**.

Example 8.5

Write down the general term of the geometric sequence 5, −10, 20, ...

$a = 5$ and $r = -2$, so $u_k = 5(-2)^{k-1} = (-1)^{k-1} \, 5 \times 2^{k-1}$.

In any geometric sequence, $r = \dfrac{u_2}{u_1} = \dfrac{u_3}{u_2} = \dots = \dfrac{u_n}{u_{n-1}}$.

This can be used to test whether a number sequence is in fact geometric, and also to find the value of r if you cannot spot it easily.

SUM OF A GEOMETRIC SERIES

DON'T FORGET

Sum (geometric)
$= \dfrac{a(1-r^n)}{1-r}$ $(r \neq 1)$

The nth partial sum is $a + ar + ar^2 + \dots + ar^{n-1} = \dfrac{a(1-r^n)}{1-r} = \dfrac{a(r^n - 1)}{r-1}$ $(r \neq 1)$.

Example 8.6

Obtain the sum of the first 20 terms of the geometric sequence 3, 1, $\frac{1}{3}$, ...

We have $a = 3$ and $r = \frac{1}{3}$, so the sum of 20 terms is:

$$\frac{3(1-(\frac{1}{3})^{20})}{1-\frac{1}{3}} = \frac{9}{2}(1 - 3^{-20}).$$

SUM TO INFINITY

When $-1 < r < 1$, $r^n \to 0$ as $n \to \infty$, and so:

$$\frac{a(1-r^n)}{1-r} \to \frac{a}{1-r}.$$

This is called the **sum to infinity**, and we write:

$$a + ar + ar^2 + \dots = \sum_{n=1}^{\infty} ar^{n-1} = \frac{a}{1-r}.$$

DON'T FORGET

The sum to infinity of a geometric series is
$\dfrac{a}{1-r}$ provided that $-1 < r < 1$.

Example 8.7

$$1 + \frac{1}{2} + \frac{1}{4} + \frac{1}{8} + \dots = \frac{1}{1-\frac{1}{2}} = 2$$

Example 8.8

For what values of x does the infinite series $1 + 2x^2 + 4x^4 + 8x^6 + \dots$ have a sum, and what is this sum?

This is a geometric series with $a = 1$ and $r = 2x^2$, so we need $-1 < 2x^2 < 1$.

The lower limit is automatically true, and the upper limit gives:

$$-\frac{1}{\sqrt{2}} < x < \frac{1}{\sqrt{2}}.$$

The sum is given by:

$$\frac{a}{1-r} = \frac{1}{1-2x^2}.$$

MORE SUMS

Example 8.9

Evaluate the sum $\displaystyle\sum_{r=10}^{20}(3r-5)$.

Method 1

$$\sum_{r=10}^{20}(3r-5) = \left(3\sum_{r=1}^{20}r - \sum_{r=1}^{20}5\right) - \left(3\sum_{r=1}^{9}r - \sum_{r=1}^{9}5\right)$$

$$= \left(\frac{3}{2}\times 20(21) - 5\times 20\right) - \left(\frac{3}{2}\times 9(10) - 6\times 9\right)$$

$$= (630-100)-(135-45)$$

$$= (530-90)$$

$$= 440.$$

DON'T FORGET

$$\sum_{n=k}^{m} = \sum_{n=1}^{m} - \sum_{n=1}^{k-1}$$

Note that

$\displaystyle\sum_{n=k}^{m} = \sum_{n=1}^{m} - \sum_{n=1}^{k-1}$ applies to all series, and you will use it when you want to find the sum of terms which start part-way along the series.

Method 2

$$\sum_{r=10}^{20}(3r-5) = \sum_{r=1}^{20}(3r-5) - \sum_{r=1}^{9}(3r-5)$$

$$= (-2+1+4+\ldots+55) - (-2+1+4+\ldots+22)$$

$$= \frac{20}{2}(-2+55) - \frac{9}{2}(-2+22) = 530-90 = 440.$$

The square and cube numbers do not form arithmetic or geometric sequences. However, the formula for the sum of their terms is known. Remember these results:

$$\sum_{k=1}^{n}k = \frac{n(n+1)}{2} \qquad \sum_{k=1}^{n}k^2 = \frac{n(n+1)(2n+1)}{6} \qquad \sum_{k=1}^{n}k^3 = \left(\frac{n(n+1)}{2}\right)^2$$

NOW TRY THIS

(1) The second and fourth terms of a geometric series are -2 and $-\frac{2}{9}$ respectively.

Explain why the series has a sum to infinity, and obtain two possible values for this sum. **5**

(2) The first three terms of a geometric sequence are

$\dfrac{x(x+1)}{(x-2)}$, $\dfrac{x(x+1)^2}{(x-2)^2}$ and $\dfrac{x(x+1)^3}{(x-2)^3}$, where $x<2$.

(a) Obtain expressions for the common ratio and the nth term of the sequence. **3**

(b) Find an expression for the sum of the first n terms of the sequence. **3**

(c) Obtain the range of values of x for which the sequence has a sum to infinity, and find an expression for the sum to infinity. **4**

(3) The first two terms of a geometric sequence are $a_1 = p$ and

$a_2 = p^2$. Obtain expressions for S_n and S_{2n} in terms of p, where $S_k = \displaystyle\sum_{j=1}^{k}a_j$. **1, 1**

Given that $S_{2n} = 65S_n$, show that $p^n = 64$. **2**

Given also that $a_3 = 2p$ and that $p > 0$, obtain the exact value of p and hence the value of n. **1, 1**

MACLAURIN SERIES

DON'T FORGET

The Maclaurin series of $f(x)$ is $f(0) + \sum_{r=1}^{\infty} \dfrac{f^{(r)}(0).x^r}{r!}$

When you use your calculator to find the value of sin 40°, how can it give an answer to such accuracy? Certainly not from any scale drawing involving a 40° triangle.

Here is one of the methods which are used to evaluate accurately advanced formulae such as trigonometric, exponential and logarithmic functions.

Suppose a function $f(x)$ has derivatives of all orders defined at $x = 0$. Denote the rth derivative evaluated at $x = 0$ by $f^{(r)}(0)$ for $r = 1, 2, 3, \ldots$, and form the infinite series:

$$f(0) + f'(0)x + \frac{f''(0)x^2}{2!} + \ldots = f(0) + \sum_{r=1}^{\infty} \frac{f^{(r)}(0)x^r}{r!}.$$

When this series converges (the values of x may be restricted for this), it is called the Maclaurin series (or expansion) of $f(x)$, after the Scottish mathematician Colin Maclaurin (1698–1746), who did much early work on these series.

The sum is $f(x)$ when the series converges for all the functions you will meet at Advanced Higher.

Example 8.10

Find the Maclaurin series for $f(x) = \dfrac{1}{1-x}$.

We have $f(0) = 1$, $f'(0) = 1$, $f''(0) = 2$, and in general $f^{(r)}(0) = r!$ (verify a few more derivatives to see the pattern).

So, the Maclaurin series for

$\dfrac{1}{1-x}$ is $1 + x + x^2 + \ldots$.

From the section on geometric series, we know that this series only converges for $-1 < x < 1$,

so although the function $f(x) = \dfrac{1}{1-x}$ is defined for all $x \neq 1$,

its Maclaurin series is only defined for $-1 < x < 1$.

Example 8.11

Obtain the Maclaurin series for $f(x) = (1 + x)^n$ where n is a positive integer.

$f'(x) = n(1 + x)^{n-1}$, $f''(x) = n(n - 1)(1 + x)^{n-2}$,   and generally:

$f^{(r)}(x) = n(n - 1) \ldots (n - r + 1)(1 + x)^{n-r}$ for $r \leqslant n$.   For $r > n$, all derivatives are zero.

So, the Maclaurin series for $(1 + x)^n$ is:

$$1 + nx + \frac{n(n-1)}{2!}x^2 + \ldots + \frac{n(n-1)\ldots(n-r+1)}{r!}x^r + \ldots + x^n$$

$$= 1 + \binom{n}{1}x + \binom{n}{2}x^2 + \ldots + \binom{n}{r}x^r + \ldots + \binom{n}{n}x^n.$$

Because this is a finite series (polynomial), it is valid for all x. We recognise this as the binomial expansion from Chapter 1. Now we can use Maclaurin series to expand binomials with fractional and negative powers.

Example 8.12

Obtain the first four terms in the Maclaurin series for $\sqrt{1+x}$.

$f(x) = \sqrt{1+x}, \; f(0) = 1$

$f'(x) = \frac{1}{2}(1+x)^{-\frac{1}{2}}, \; f'(0) = \frac{1}{2}$

$f''(x) = \left(\frac{1}{2}\right)\left(-\frac{1}{2}\right)(1+x)^{-\frac{3}{2}}, \; f''(0) = -\frac{1}{4}$

$f'''(x) = \left(\frac{1}{2}\right)\left(-\frac{1}{2}\right)\left(-\frac{3}{2}\right)(1+x)^{-\frac{5}{2}}, \; f'''(0) = \frac{3}{8}.$

So, for the first four terms, we have:

$$\sqrt{1+x} = 1 + \frac{x}{2} - \frac{x^2}{8} + \frac{x^3}{16} + \ldots$$

The series converges for $-1 < x < 1$ (you are not expected to prove this), although the function is defined for all $x > -1$.

The following table gives some standard results which are worth remembering. You are expected to know how to derive them.

Function	Maclaurin series	Valid for	Function	Maclaurin series	Valid for
e^x	$1 + x + \frac{x^2}{2!} + \frac{x^3}{3!} + \ldots$	all x	$\cos x$	$1 - \frac{x^2}{2!} + \frac{x^4}{4!} - \ldots$	all x
$\sin x$	$x - \frac{x^3}{3!} + \frac{x^5}{5!} - \ldots$	all x	$\ln(1+x)$	$x - \frac{x^2}{2} + \frac{x^3}{3} - \ldots$	$-1 < x \leqslant 1$

DON'T FORGET

Try to use standard results as much as possible.

LET'S THINK ABOUT THIS

Why does the Maclaurin series for $\cos x$ only contain even powers, while that for $\sin x$ only contains odd powers?

Answer: $\cos x$ is an even function of x, and $\sin x$ is an odd function of x.

ASSOCIATED SERIES

Using the above table, we can obtain Maclaurin expansions for more complicated functions.

Example 8.13

Obtain the Maclaurin series for $\sin^2 x$ as far as the term in x^4. Note that, if a question hasn't directed you to use a particular method, any of these methods would be acceptable.

DON'T FORGET

Be prepared to combine Maclaurin series with other algebra or trigonometry topics.

Method 1

$\sin^2 x = \frac{1}{2}(1 - \cos 2x)$

$= \frac{1}{2} - \frac{1}{2}\left(1 - \frac{(2x)^2}{2!} + \frac{(2x)^4}{4!} - \ldots\right)$

$= \frac{1}{2} - \frac{1}{2} + \frac{4x^2}{4} - \frac{16x^4}{48} + \ldots$

$= x^2 - \frac{x^4}{3} + \ldots$

Method 2

$f(x) = \sin^2 x, \; f(0) = 0$

$f'(x) = 2\sin x \cos x = \sin 2x, \; f'(0) = 0$

Using the double angle formula here means that we do not require the product rule for further derivatives.

$f''(x) = 2\cos 2x, \quad f''(0) = 2$

$f'''(x) = -4\sin 2x, \quad f'''(0) = 0$

$f''''(x) = -8\cos 2x, \quad f''''(0) = -8$

Check that these results give the same answer as before.

Method 3

$\sin^2 x = \sin x \sin x = \left(x - \frac{x^3}{6} + \ldots\right)\left(x - \frac{x^3}{6} + \ldots\right)$

(ignoring higher powers)

$= x^2 - \frac{x^4}{6} - \frac{x^4}{6} + \ldots$, giving the same result as before.

contd

SEQUENCES AND SERIES

Example 8.14

Obtain the Maclaurin series for $f(x) = xe^{-x^2}$ as far as the term in x^5.

Replacing x by $-x^2$ in the expansion for e^x gives:

$$e^{-x^2} = 1 - x^2 + \frac{x^4}{2} - \dots, \text{ so:}$$

$$f(x) = x - x^3 + \frac{x^5}{2} - \dots$$

To differentiate the function xe^{-x^2} five times using the product and chain rules would involve far more work than the marks awarded (probably about 3 or 4) would justify.

NOW TRY THIS

(1) Obtain the Maclaurin series for $\sin x$ as far as the term in x^5.   **2**

     Deduce the Maclaurin series for $\sin 3x$ as far as the term in x^5.   **2**

     Hence obtain the Maclaurin series for $(1 + x)\sin 3x$ as far as the term in x^5.   **1**

(2) Obtain the first three non-zero terms in the Maclaurin series of $\ln(3 - x)$.   **3**

     Hence, or otherwise, obtain the first three non-zero terms in the Maclaurin series of $x^2 \ln(3 - x)$ and $x^2 \ln(3 + x)$.   **2**

     Hence obtain the first **two** non-zero terms in the Maclaurin expansion of $x^2 \ln(9 - x^2)$.   **2**

(3) Express $f(x) = \dfrac{x^2 + 8x + 11}{(x + 1)(x + 3)^2}$ $(x \neq -1, -3)$ in partial fractions.   **4**

     Hence obtain the first three non-zero terms in the Maclaurin expansion of $f(x)$.

     For what values of x does this series converge?   **5**

(4) Write down the Maclaurin expansion of e^x as far as the term in x^4.   **2**

     Deduce the Maclaurin expansion of e^{x^2} as far as the term in x^4.   **1**

     Hence, or otherwise, find the Maclaurin expansion of e^{x+x^2} as far as the term in x^4.   **3**

ITERATIVE SCHEMES

Consider the sequence $\{x_n\}$ defined by $x_{n+1} = \frac{2}{3}x_n + \frac{1}{x_n^2}$, with $x_1 = 1\cdot5$. The second term is $x_2 = \frac{2}{3} \times \frac{3}{2} + \frac{4}{9} = 1\cdot44444$ to 5 decimal places (in problems like these, it is better to use decimal approximations).

So, $x_3 = \frac{2}{3} \times 1\cdot44444 + \frac{1}{1\cdot4444^2} = 1\cdot44225$ to 5 decimal places,

and similarly (check the arithmetic) $x_4 = 1\cdot44225$ to 5 decimal places.

It seems as if the sequence is tending to a limit L which is approximately $1\cdot44225$.

If the sequence does tend to a limit L, then, because $x_n \to L$ and $x_{n+1} \to L$ as $n \to \infty$, L must satisfy the equation $L = \frac{2}{3}L + \frac{1}{L^2}$, which rearranges to $L^3 = 3$.

The generated sequence appears to (and does) converge to $\sqrt[3]{3}$, and the third term gives $\sqrt[3]{3}$ accurate to 5 decimal places. (Not all such iterative schemes converge so well – or at all.)

Example 8.15

The sequence $x_{n+1} = \frac{3}{x_n^2}$, with $x_1 = 1\cdot5$, looks as if it should converge to $\sqrt[3]{3}$, but the first few terms show no signs of converging: rounding to 5 decimal places,
$x_2 = 1\cdot33333$, $x_3 = 1\cdot68750$, $x_4 = 1\cdot05350$, $x_5 = 2\cdot70305$.

Any equation $f(x) = 0$ can always be rearranged (sometimes in several ways) to an equation of the form $x = g(x)$, from which we can define a sequence $x_{n+1} = g(x_n)$ $(n = 1, 2, ...)$, with x_1 taken as a first approximation to a root of $f(x) = 0$. Iterative sequences are used to solve equations for which we don't have simple solution formulae – these are in fact the majority of equations.

The iterative scheme described above can be illustrated geometrically using a cobweb diagram.

 www.waldomaths.com/Cobweb1NL.jsp

CONVERGENCE

How do we know which sequences will converge? The answer is to look at $g'(x)$:

If $|g'(x)| < 1$ in an interval containing at least one root of $L = g(L)$, and the initial guess x_1 lies in this interval, then the sequence $x_{n+1} = g(x_n)$ $(n = 1, 2, ...)$ will converge to a limit L satisfying $L = g(L)$.

L is called a **fixed point** of the iterative scheme. Fixed points of an iterative scheme $x_{n+1} = g(x_n)$ $(n = 1, 2, ...)$ can be found algebraically.

DON'T FORGET

L is a fixed point of the iterative scheme $x_{n+1} = g(x_n)$ $(n = 1, 2, ...)$ if it satisfies the equation $L = g(L)$.

 www.ugrad.math.ubc.ca/coursedoc/math101/notes/series/convergence.html

NOW TRY THIS

(1) Obtain algebraically the fixed point of the iterative scheme given by
$x_{n+1} = \ln(1 + \frac{1}{2}e^{x_n})$ $(n = 1, 2, ...)$.   **3**

(2) A recurrence relation is defined by the formula: $x_{n+1} = \frac{1}{2}\left\{x_n + \frac{7}{x_n}\right\}$.

Find the fixed points of this recurrence relation.   **3**

BASIC SKILLS INFORMATION

DON'T FORGET

Distinguish between:
- vectors   u
- scalars   a

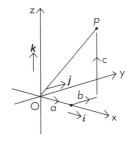

Vectors have magnitude and direction. Examples are velocity, force and displacements in three-dimensional space. Numbers are called **scalars** to distinguish them from vectors.

We denote vectors by boldface type, e.g. **u**, but in your written work you should underline vectors, e.g. u.

Unit vectors have magnitude 1. The unit displacement vectors along the positive Ox-, Oy- and Oz-axes in three-dimensional space are denoted by **i**, **j** and **k**.

For a point P with coordinates (a, b, c), the position vector $\overrightarrow{OP}$ is the displacement from O to P $a\boldsymbol{i} + b\boldsymbol{j} + c\boldsymbol{k}$, where '+' means a units along the x-axis, followed by b units along the y-axis, then c units along the z-axis.

Another way to represent the vector $a\boldsymbol{i} + b\boldsymbol{j} + c\boldsymbol{k}$ is the column matrix $\begin{pmatrix} a \\ b \\ c \end{pmatrix}$.

This is called a **column vector**.

The scalars a, b and c are called the **components** of the vector $a\boldsymbol{i} + b\boldsymbol{j} + c\boldsymbol{k}$.

The vector $\begin{pmatrix} 0 \\ 0 \\ 0 \end{pmatrix}$ is the **zero vector**, denoted by **0**.

If $\boldsymbol{p} = a\boldsymbol{i} + b\boldsymbol{j} + c\boldsymbol{k}$, then for any scalar k:

$k\boldsymbol{p} = ka\boldsymbol{i} + kb\boldsymbol{j} + kc\boldsymbol{k}$.

The magnitude of the vector $a\boldsymbol{i} + b\boldsymbol{j} + c\boldsymbol{k}$ is $\sqrt{a^2 + b^2 + c^2}$ (length and magnitude are always ≥ 0), so the unit vectors pointing in the direction determined by $\overrightarrow{OP}$ are given by:

$$+\frac{a\boldsymbol{i} + b\boldsymbol{j} + c\boldsymbol{k}}{\sqrt{a^2 + b^2 + c^2}} \quad \text{and} \quad -\frac{a\boldsymbol{i} + b\boldsymbol{j} + c\boldsymbol{k}}{\sqrt{a^2 + b^2 + c^2}}.$$

The minus sign gives the unit vector in the direction $\overrightarrow{PO}$.

Addition is defined by:

$$(a_1\boldsymbol{i} + b_1\boldsymbol{j} + c_1\boldsymbol{k}) + (a_2\boldsymbol{i} + b_2\boldsymbol{j} + c_2\boldsymbol{k}) = (a_1 + a_2)\boldsymbol{i} + (b_1 + b_2)\boldsymbol{j} + (c_1 + c_2)\boldsymbol{k}.$$

Example 9.1

If $\boldsymbol{a} = 2\boldsymbol{i} + \boldsymbol{j} - 3\boldsymbol{k}$ and $\boldsymbol{b} = \boldsymbol{i} - 2\boldsymbol{j} + \boldsymbol{k}$,   obtain $3\boldsymbol{a} - 2\boldsymbol{b}$.

$3\boldsymbol{a} - 2\boldsymbol{b} \quad = \quad 3(2\boldsymbol{i} + \boldsymbol{j} - 3\boldsymbol{k}) - 2(\boldsymbol{i} - 2\boldsymbol{j} + \boldsymbol{k}) \quad = \quad 4\boldsymbol{i} + 7\boldsymbol{j} - 11\boldsymbol{k}$.

SCALAR PRODUCT

Given two vectors **a** and **b**, we define the **scalar product** of **a** and **b** by:

$\boldsymbol{a}.\boldsymbol{b} = |\boldsymbol{a}|\,|\boldsymbol{b}|\cos\theta$.

Note that:

$\boldsymbol{a}.\boldsymbol{b} = \boldsymbol{b}.\boldsymbol{a}$ and $\boldsymbol{a}.(\boldsymbol{b} + \boldsymbol{c}) = \boldsymbol{a}.\boldsymbol{b} + \boldsymbol{a}.\boldsymbol{c}$.

When $\theta = 90°$, $\cos\theta = 0$, so:

$\boldsymbol{a}.\boldsymbol{b} = 0$ and **a** and **b** are perpendicular.

We have $\boldsymbol{i}.\boldsymbol{j} = \boldsymbol{j}.\boldsymbol{k} = \boldsymbol{k}.\boldsymbol{i} = 0$, and $\boldsymbol{i}.\boldsymbol{i} = \boldsymbol{j}.\boldsymbol{j} = \boldsymbol{k}.\boldsymbol{k} = 1$.

Using these results, we have:

$$(a_1\boldsymbol{i} + b_1\boldsymbol{j} + c_1\boldsymbol{k}).(a_2\boldsymbol{i} + b_2\boldsymbol{j} + c_2\boldsymbol{k}) = a_1a_2 + b_1b_2 + c_1c_2.$$

DON'T FORGET

The scalar product of two vectors is a number, not a vector.

contd

SCALAR PRODUCT   contd

Example 9.2

For what value of a are the vectors $2i + j + k$ and $i + aj - 3k$ perpendicular to each other?

$(2i + j + k).(i + aj - 3k)$ $=$ $2 + a - 3$ $=$ 0, so $a = 1$.

VECTOR PRODUCT

As well as the scalar product, we can also define a product of two vectors which gives a vector: $a \times b$ (read as 'a cross b') is a vector with magnitude $|a||b| \sin\theta$ pointing as shown in the diagram.

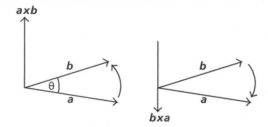

Note that $a \times b = -b \times a$ and $a \times a = 0$.
Also $a \times (b + c) = a \times b + a \times c$.

We have $i \times j = k$, $j \times k = i$ and $k \times i = j$.

Applying the above rules, we get:

$(a_1 i + b_1 j + c_1 k) \times (a_2 i + b_2 j + c_2 k)$ $=$ $(b_1 c_2 - b_2 c_1)i - (a_1 c_2 - a_2 c_1)j + (a_1 b_2 - a_2 b_1)k.$

This is best remembered as a determinant: $\begin{vmatrix} i & j & k \\ a_1 & b_1 & c_1 \\ a_2 & b_2 & c_2 \end{vmatrix}$.

The two types of product can be combined in various ways.

> **DON'T FORGET**
> - Scalar product, . , = Scalar.
> - Vector product, × , = Vector.

Example 9.3

If $a = i + 2j - k$, $b = 2i - j + 2k$ and $c = i + j$, obtain $a.(b \times c)$ and $a \times (b \times c)$.

$$b \times c = \begin{vmatrix} i & j & k \\ 2 & -1 & 2 \\ 1 & 1 & 0 \end{vmatrix}$$

$$= i \begin{vmatrix} -1 & 2 \\ 1 & 0 \end{vmatrix} - j \begin{vmatrix} 2 & 2 \\ 1 & 0 \end{vmatrix} + k \begin{vmatrix} 2 & -1 \\ 1 & 1 \end{vmatrix} = -2i + 2j + 3k.$$

So:

$a.(b \times c) = (i + 2j - k).(-2i + 2j + 3k) = -2 + 4 - 3 = -1.$

$$a \times (b \times c) = \begin{vmatrix} i & j & k \\ 1 & 2 & -1 \\ -2 & 2 & 3 \end{vmatrix}$$

$$= i \begin{vmatrix} 2 & -1 \\ 2 & 3 \end{vmatrix} - j \begin{vmatrix} 1 & -1 \\ -2 & 3 \end{vmatrix} + k \begin{vmatrix} 1 & 2 \\ -2 & 2 \end{vmatrix}$$

$$= 8i - j + 6k.$$

NOW TRY THIS

(1) Given $u = i + j - k$, $v = 2i - j + 2k$ and $w = -i + 2k$, calculate $u.(v \times w)$.

4

EQUATIONS OF LINES

EQUATION OF A LINE

A line L in space is determined by its direction vector, d, and any point Q on the line.

For any point P on L, we have $\overrightarrow{QP} = \lambda d$ for some scalar λ, and as λ varies from $-\infty$ to $+\infty$ we get all points on L.

If A is a point on L with position vector a, and P has position vector r, then $r = a + \lambda d$.

This is the **vector equation** of a line.

> **At Higher:**
>
> For the equation of a line in two dimensions, you needed:
>
> - a gradient and a point, or
> - two points.
>
> **At Advanced Higher:**
>
> For the equation of a line in three dimensions, you need:
>
> - a vector and a point, or
> - two points.

> **DON'T FORGET**
>
> There are three forms for the equation of a line:
>
> Vector: $\qquad r = a + \lambda d$
>
> Parametric: $\begin{array}{l} x = a + \lambda d_1 \\ y = b + \lambda d_2 \\ z = c + \lambda d_3 \end{array}$
>
> Symmetric:
> $$\frac{x-a}{d_1}$$
> $$= \frac{y-b}{d_2}$$
> $$= \frac{z-c}{d_3} = \lambda$$

If $r = xi + yj + zk$, and $q = ai + bj + ck$, and $d = d_1i + d_2j + d_3k$, then equating the i, j and k components gives the **parametric equations** of a line:

$$x = a + \lambda d_1 \qquad y = b + \lambda d_2 \qquad z = c + \lambda d_3.$$

An equivalent form to these equations are the **symmetric** (or **Cartesian**) **equations**:

$$\frac{x-a}{d_1} = \frac{y-b}{d_2} = \frac{z-c}{d_3} = \lambda$$

Although the equation is often not given in exam questions with the parameter shown, you should always include it in your working.

Example 9.4

Obtain all three forms for the equation of the line which passes through the points $A(1, -1, 3)$ and $B(3, 0, 1)$.

We need the direction vector d.

This is: $\qquad \overrightarrow{AB} = \begin{pmatrix} 3 \\ 0 \\ 1 \end{pmatrix} - \begin{pmatrix} 1 \\ -1 \\ 3 \end{pmatrix} = \begin{pmatrix} 2 \\ 1 \\ -2 \end{pmatrix}$,

so: $\qquad d = 2i + j - 2k$.

The vector equation is given by:

$$xi + yj + zk \quad = \quad i - j + 3k + \lambda(2i + j - 2k).$$

(Either point, A or B, may be used to get the equations.)

The parametric equations using A are:

$$x = 1 + 2\lambda, \quad y = -1 + \lambda, \quad z = 3 - 2\lambda.$$

The symmetric form is given by:

$$\frac{x-1}{2} = \frac{y+1}{1} = \frac{z-3}{-2} = \lambda.$$

contd

EQUATION OF A LINE contd

Note that, if we use B, we get the parametric equations:

$$x = 3 + 2\mu, \quad y = \mu, \quad z = 1 - 2\mu.$$

Although these look quite different, they describe the same line. For example, $\mu = -1$ gives the point A.

DISTANCE OF A POINT FROM A LINE

Given a point P and a line L, the distance of P from L is the length PQ, where Q is the point on L such that $\overrightarrow{PQ}$ is perpendicular to the direction of L.

Example 9.5

Obtain the distance of the point $P(1, 1, 1)$ from the line L with parametric equations:

$$x = 2 + \lambda, \quad y = 2 - \lambda, \quad z = 1 + 2\lambda.$$

Q has coordinates $(2 + \lambda, 2 - \lambda, 1 + 2\lambda)$ for a value of λ to be determined.

Hence $\overrightarrow{PQ} = \begin{pmatrix} 2 + \lambda \\ 2 - \lambda \\ 1 + 2\lambda \end{pmatrix} - \begin{pmatrix} 1 \\ 1 \\ 1 \end{pmatrix} = \begin{pmatrix} 1 + \lambda \\ 1 - \lambda \\ 2\lambda \end{pmatrix}$

The direction vector of L is $\boldsymbol{d} = \begin{pmatrix} 1 \\ -1 \\ 2 \end{pmatrix}$, and we require $\overrightarrow{PQ} \cdot \boldsymbol{d} = 0$.

This gives $(1 + \lambda) - (1 - \lambda) + 4\lambda = 0$, so $\lambda = 0$.

So, Q has coordinates $(2, 2, 1)$ and:

$$PQ = \sqrt{(2 - 1)^2 + (2 - 1)^2 + (1 - 1)^2} = \sqrt{2}.$$

So, the distance of P from L is $\sqrt{2}$.

NOW TRY THIS

(1) Lines L_1 and L_2 are given by the equations

$$L_1\colon \frac{x+3}{-2} = \frac{y-1}{1} = \frac{z-5}{3} \qquad L_2\colon \frac{x-1}{1} = \frac{y-1}{-1} = \frac{z-3}{-1}.$$

   (a) Show that L_1 and L_2 do not intersect.   **3**

   (b) Obtain parametric equations for the line L_3 which passes through $P(-1, 0, 2)$ and has direction vector perpendicular to the directions of L_1 and L_2.   **3**

   (c) Show that L_3 intersects L_2 and obtain the point of intersection Q.   **3**

   (d) Verify that P lies on L_1 and explain why the distance PQ gives the shortest distance between L_1 and L_2.   **1**

EQUATION OF A PLANE

To obtain an equation of a plane, we need one of the following:

- three points on the plane

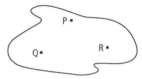

- a point on the plane and a direction vector normal to the plane

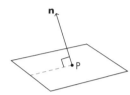

- two non-parallel lines lying in the plane.

DON'T FORGET

$r.n = a.n$

From the second of the figures, we have $(r - a).n = 0$, where r is the position vector of a general point on the plane, a is the position vector of a fixed point on the plane and n is a vector normal to the plane.

So, if $r = xi + yj + zk$, $a = a_1i + a_2j + a_3k$ and $n = n_1i + n_2j + n_3k$, we get:

$n_1(x - a_1) + n_2(y - a_2) + n_3(z - a_3) = 0$, which leads to:

$$n_1x + n_2y + n_3z = n_1a_1 + n_2a_2 + n_3a_3.$$

This has the form $n_1x + n_2y + n_3z = d$, where $\begin{pmatrix} n_1 \\ n_2 \\ n_3 \end{pmatrix}$ is a normal vector to the plane and d is a

constant. This is the standard form of the equation of a plane.

Example 9.6

Obtain an equation for the plane passing through the points $P(1, 1, 2)$, $Q(1, -1, 1)$ and $R(2, 4, 2)$.

Method 1

The equation of the plane has the form $ax + by + cz = d$, so substituting the coordinates for P, Q and R gives the equations:

$$a + b + 2c = d, \quad a - b + c = d, \quad 2a + 4b + 2c = d.$$

Solving these for a, b and c using the methods of Chapter 5 gives:

$$a = \frac{d}{2}, \quad b = \frac{-d}{6} \quad \text{and} \quad c = \frac{d}{3}.$$

Substituting into the equation $ax + by + cz = d$ and simplifying gives the equation $3x - y + 2z = 6$.

Method 2

The vector $\overrightarrow{PQ} \times \overrightarrow{PR}$ is normal to the plane.

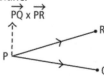

$$\overrightarrow{PQ} = \overrightarrow{OQ} - \overrightarrow{OP} = \begin{pmatrix} 1 \\ -1 \\ 1 \end{pmatrix} - \begin{pmatrix} 1 \\ 1 \\ 2 \end{pmatrix} = \begin{pmatrix} 0 \\ -2 \\ -1 \end{pmatrix}$$

$$\overrightarrow{PR} = \overrightarrow{OR} - \overrightarrow{OP} = \begin{pmatrix} 2 \\ 4 \\ 2 \end{pmatrix} - \begin{pmatrix} 1 \\ 1 \\ 2 \end{pmatrix} = \begin{pmatrix} 1 \\ 3 \\ 0 \end{pmatrix}$$

$$\overrightarrow{PQ} \times \overrightarrow{PR} = \begin{vmatrix} i & j & k \\ 0 & -2 & -1 \\ 1 & 3 & 0 \end{vmatrix} = 3i - j + 2k.$$

So, a vector normal to the plane is:

$$3i - j + 2k, \qquad n = \begin{pmatrix} 3 \\ -1 \\ 2 \end{pmatrix},$$

and the equation of the plane is of the form $3x - y + 2z = d$.

Substituting the coordinates of P: $(3(1) - (1) + 2(2) = 6)$ gives $d = 6$.
As a check, use the coordinates of Q or R to give this same d value.

So, the equation of the plane is $3x - y + 2z = 6$.

Example 9.7

Obtain an equation for the plane containing the lines:

$$L_1: x = 1 + 2\lambda, \qquad y = 2 - \lambda, \qquad z = 1 - 2\lambda,$$
$$L_2: x = 1 - \mu, \qquad y = 2 + \mu, \qquad z = 1 + 3\mu.$$

We first check that the lines intersect: setting the x-coordinates of the two lines equal, and also the y-coordinates, gives

$$1 + 2\lambda = 1 - \mu \qquad \text{and} \qquad 2 - \lambda = 2 + \mu, \qquad \text{so} \qquad \lambda = \mu = 0.$$

These values also make the z-coordinates equal, so the two lines intersect at the point $P(1, 2, 1)$.

The direction of the normal to the plane containing L_1 and L_2 is given by:

$$\begin{vmatrix} i & j & k \\ 2 & -1 & -2 \\ -1 & 1 & 3 \end{vmatrix} = i\begin{vmatrix} -1 & -2 \\ 1 & 3 \end{vmatrix} - j\begin{vmatrix} 2 & -2 \\ -1 & 3 \end{vmatrix} + k\begin{vmatrix} 2 & -1 \\ -1 & 1 \end{vmatrix}$$

$$= -i - 4j + k$$

so, the equation of the plane has the form:

$$-x - 4y + z = d.$$

Substituting the intersection point $x = 1$, $y = 2$, $z = 1$ gives $d = -8$.

So (dividing through by -1), an equation for the plane is:

$$x + 4y - z = 8.$$

INTERSECTIONS OF LINES AND PLANES

INTERSECTION OF TWO LINES

When solving this type of problem, it is essential to use different parameters for the two different lines.

> **DON'T FORGET**
>
> Use different parameters for two different lines.

Example 9.8

Lines L_1 and L_2 are defined as follows:

L_1: $x = 3 + \lambda$, $y = 4 + 2\lambda$, $z = -\lambda$

L_2: $x = 3 + 2\mu$, $y = 3 + 3\mu$, $z = 4 + 2\mu$.

Decide if they intersect – and, if they do, obtain the point of intersection.

The lines intersect if the following set of three equations in two unknowns has a solution:

$3 + \lambda = 3 + 2\mu$, $4 + 2\lambda = 3 + 3\mu$, $-\lambda = 4 + 2\mu$.

These equations are formed by equating x from L_1 with x from L_2 etc.

> **DON'T FORGET**
>
> When checking if two lines meet, all three equations for the parameters must be checked.

The procedure is to solve any pair of equations and **check that this solution satisfies the third equation**. If it does, the value of λ or μ will give the point of intersection. If it doesn't, the lines do not intersect.

Solving the first two equations gives $\lambda = -2$ and $\mu = -1$, and these satisfy the third equation, so the lines intersect.

Putting $\lambda = -2$ in the equations for L_1 (or $\mu = -1$ in L_2) gives the intersection point $(1, 0, 2)$.

INTERSECTION OF A LINE AND PLANE

The best way to solve this type of problem is to use the standard equation of the plane and a parametric equation of the line.

Example 9.9

Obtain the point of intersection of the line given by $x = 2 - \lambda$, $y = -2\lambda$, $z = 1 + 2\lambda$ and the plane with equation $x + 2y + z = 0$.

All we need to do is substitute the parametric equations of the line into the equation of the plane, so:

$(2 - \lambda) + 2(-2\lambda) + (1 + 2\lambda) = 0$.

Solving gives $\lambda = 1$, and the point of intersection is given by setting $\lambda = 1$ in the parametric equations of the line. This gives the point of intersection as $(1, -2, 3)$.

Two other situations can arise:

- the line is parallel to the plane
- the line lies completely in the plane.

Example 9.10

Show that, for all values of a except one, the line given by:

$x = a + \lambda, y = 2 + \lambda, z = 1 - \lambda$

does not intersect the plane with the equation:

$2x - y + z = 1$.

What happens for this exceptional value of a?

contd

INTERSECTION OF A LINE AND PLANE   contd

Substituting the parametric equations into the equation of the plane gives:

$$2(a + \lambda) - (2 + \lambda) + (1 - \lambda) = 1 \implies 2a - 1 = 1.$$

Note that the λ terms cancel out, and for all $a \neq 1$ we are left with an impossibility.

Hence, for all values of a except 1, the line does not intersect the plane (they are parallel).

When $a = 1$, we get the identity $1 = 1$. This means that the coordinates of every point on the line satisfy the condition $2x - y + z = 1$ automatically, i.e. the line lies completely in the plane.

DISTANCE OF A POINT FROM A PLANE

Given a point P and a plane π, the distance of P from π is the length PQ, where Q is the point on π such that $\overrightarrow{PQ}$ is perpendicular to π.

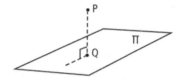

Example 9.11

Obtain the distance of the point $P(2, -1, 3)$ from the plane π with equation:

$$2x + y - z = 6.$$

Let Q be the point on π such that $\overrightarrow{PQ}$ is perpendicular to π, which means that the direction vector of the line PQ is $2\mathbf{i} + \mathbf{j} - \mathbf{k}$ (here the direction vector of the line is obviously the same as the normal vector to the plane).

Hence the parametric equations of the line PQ are:

$$x = 2 + 2\lambda \qquad y = -1 + \lambda \qquad z = 3 - \lambda.$$

This line meets π when $2(2 + 2\lambda) + (-1 + \lambda) - (3 - \lambda) = 6$, so $\lambda = 1$.

Putting $\lambda = 1$ in the parametric equations gives $Q(4, 0, 2)$.

The distance of P from π is thus $\sqrt{(2-4)^2 + (-1-0)^2 + (3-2)^2} = \sqrt{6}$.

NOW TRY THIS

(1) (a) Obtain an equation for the plane passing through the point $A(2, 1, -1)$ which is perpendicular to the line L given by   3

$$\frac{x + 3}{1} = \frac{y}{2} = \frac{z - 12}{1}.$$

   (b) Obtain the coordinates of the point B where the plane and L intersect.   4

   (c) Calculate the distance AB and explain why this gives the shortest distance from A to L.   3

INTERSECTIONS OF PLANES

INTERSECTION OF TWO PLANES

Unless the planes are parallel, two planes intersect in a line.

Example 9.12

Obtain parametric equations for the line of intersection of the planes with equations:

$$x + 2y - z = 1 \quad \text{and} \quad 2x - y + 2z = 2.$$

Let $z = t$, where t is a parameter, and solve the equations for x and y:

$$x + 2y = 1 + t, \quad 2x - y = 2 - 2t.$$

Solving for x and y gives:

$$x = 1 - \frac{3}{5}t, \quad y = \frac{4}{5}t, \quad z = t.$$

As t is a parameter, let $t = 5s$ to get rid of fractions.
This gives the parametric equations for the line of intersection as:

$$x = 1 - 3s, \quad y = 4s, \quad z = 5s.$$

INTERSECTION OF THREE PLANES

Unique point of intersection

The unique point of intersection corresponds to the diagram shown here. Two of the planes intersect in a line, which intersects the third plane at a unique point. It corresponds to the situation where three simultaneous equations have a unique solution.

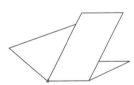

> **DON'T FORGET**
>
> There are several cases that can arise when three planes intersect:
>
> - unique point of intersection
> - line of intersection
> - no intersection.

Example 9.13

Obtain the point of intersection of the three planes given by:

$$x + y - 2z = 1 \quad x + 2y + z = 6 \quad 2x - y - z = -1.$$

Using the Gaussian elimination methods given in Chapter 5, the (unique) solution is found to be:

$$x = 1, y = 2, z = 1.$$

Hence the three planes all meet at the point with coordinates (1, 2, 1).

Line of intersection

The line of intersection corresponds to the diagram shown here. The third plane contains the line of intersection of the other two planes. It corresponds to the situation where three simultaneous equations have infinitely many solutions.

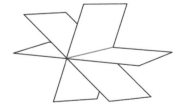

Example 9.14

Show that the planes with equations:

$$x + y + 2z = 1 \quad x - y + z = 2 \quad 3x + y + 5z = 4$$

intersect in a line, giving parametric equations for this line.

Using the Gaussian elimination methods in Chapter 5, we get:

contd

INTERSECTION OF THREE PLANES contd

$$\begin{array}{ccc|c} 1 & 1 & 2 & 1 \\ 1 & -1 & 1 & 2 \\ 3 & 1 & 5 & 4 \end{array}$$

becomes:

$$\begin{array}{ccc|cl} 1 & 1 & 2 & 1 & \\ 0 & -2 & -1 & 1 & R_2 - R_1 \\ 0 & -2 & -1 & 1 & R_3 - 3R_1 \end{array}$$

Now the last row becomes: $\qquad 0 \quad 0 \quad 0 \mid 0$

so we set $z = t$, where t is a parameter.

Solving for y, then x, we get:

$$x = \frac{3}{2} - \frac{3}{2}t, \qquad y = -\frac{1}{2} - \frac{1}{2}t, \qquad z = t.$$

These are parametric equations of a line through the point $\left(\frac{3}{2}, -\frac{1}{2}, 0\right)$ with direction vector

parallel to $\begin{pmatrix} 3 \\ 1 \\ -2 \end{pmatrix}$, so the three planes meet in a line.

No intersection

The situation when there is no intersection between three planes corresponds to one of the diagrams shown here.

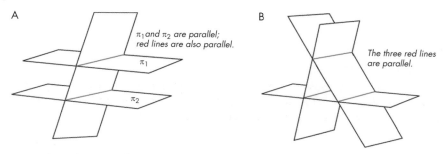

A

π_1 and π_2 are parallel;
red lines are also parallel.

π_1

π_2

B

The three red lines are parallel.

It corresponds to the case where the simultaneous equations are inconsistent.

Example 9.15

Show that the planes with equations:

$$x + y + z = 1 \qquad 2x + y - z = 2 \qquad 4x + 3y + z = 1$$

do not meet at a single point. Obtain equations for the lines of intersection:

$$\begin{array}{ccc|c} 1 & 1 & 1 & 1 \\ 2 & 1 & -1 & 2 \\ 4 & 3 & 1 & 1 \end{array}$$

becomes:

$$\begin{array}{ccc|cl} 1 & 1 & 1 & 1 & \\ 0 & -1 & -3 & 0 & R_2 - 2R_1 \\ 0 & -1 & -3 & -3 & R_3 - 4R_1 \end{array}$$

Now $R_3 - R_2$ gives the last row as: $\qquad 0 \quad 0 \quad 0 \mid -3$

so the three equations are inconsistent.

This means the three planes do not meet at a single point.

Setting $z = t$ in all three equations and solving them in pairs gives equations of the lines of intersection of the planes taken in pairs. The details are left as an exercise, giving:

- planes 1 and 2: line $x = 1 + 2t,$ $y = -3t,$ $z = t$
- planes 1 and 3: line $x = -2 + 2t,$ $y = 3 - 3t,$ $z = t$
- planes 2 and 3: line $x = \frac{5}{2} + 2t,$ $y = -3 - 3t,$ $z = t$

This example illustrates the right-hand diagram (B) above.

ANGLES BETWEEN LINES AND PLANES

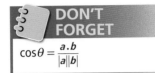

DON'T FORGET

$\cos\theta = \dfrac{a.b}{|a||b|}$

Given two directions determined by vectors a and b, we have $a.b = |a||b|\cos\theta$.

So, $\cos\theta = \dfrac{a.b}{|a||b|}$, and this gives the angle between two vectors.

ANGLE BETWEEN TWO LINES

Remember that lines in space may not intersect, so it only makes sense to calculate the angle between intersecting lines.

Example 9.16

Calculate the acute angle between the lines:

$$L_1: \quad x = 1 + t, \qquad y = 2t, \qquad z = 2 - t$$
$$L_2: \quad x = 3 + 2s, \qquad y = 3 + 3s, \qquad z = 4 + 2s.$$

Direction vectors for the lines are:

$$d_1 = i + 2j - k \qquad \text{and} \qquad d_2 = 2i + 3j + 2k$$

$$|d_1| = \sqrt{1+4+1} = \sqrt{6} \qquad \text{and} \qquad |d_2| = \sqrt{4+9+4} = \sqrt{17}$$

$$d_1 . d_2 = 2 + 6 - 2 = 6.$$

The angle between the lines is given by:

$$\cos\theta \;=\; \frac{6}{\sqrt{6}\sqrt{17}} \;=\; \frac{\sqrt{6}}{\sqrt{17}}, \qquad \text{so:}$$

$$\theta \;=\; \cos^{-1}\left(\frac{\sqrt{6}}{\sqrt{17}}\right) \;=\; 53 \cdot 6°.$$

Unless told otherwise, 3 significant figures is sufficient.

ANGLE BETWEEN A LINE AND A PLANE

Assume that the line is not parallel to the plane. Because different lines in the plane will make different angles with this intersecting line, we calculate the acute angle $\theta°$ between the intersecting line and the normal to the plane, and the angle between the line and the plane is $(90 - \theta)°$.

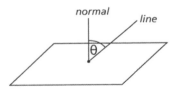

Example 9.17

Calculate the acute angle between the line given by:

$$x = 2 - t, \quad y = -2t, \quad z = 1 + 2t$$

and the plane with equation:

$$x + y + z = 0.$$

contd

ANGLE BETWEEN A LINE AND A PLANE   contd

A direction vector for the line is $d = \begin{pmatrix} -1 \\ -2 \\ 2 \end{pmatrix}$, and a normal vector for the plane is $n = \begin{pmatrix} 1 \\ 1 \\ 1 \end{pmatrix}$, so:

$$\cos\theta = \frac{-1-2+2}{\sqrt{1+1+1} \times \sqrt{1+4+4}} = \frac{-1}{3\sqrt{3}}$$

Hence $\theta = 101\cdot1°$.

The acute angle is $180° - 101\cdot1° = 78\cdot9°$, which is the angle between the line and the normal to the plane, so the angle between the line and plane is $90° - 78\cdot9° = 11\cdot1°$.

ANGLE BETWEEN TWO PLANES

The angle between two planes is the angle between their normal vectors.

Example 9.18

For what value of α do the planes with equations

$$x + 2y - z = 1 \quad \text{and} \quad \alpha x - y + 3z = 2$$

intersect at right angles?

If the angle between the planes, θ, is 90°, then $\cos\theta = 0$.

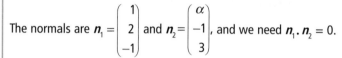

The normals are $n_1 = \begin{pmatrix} 1 \\ 2 \\ -1 \end{pmatrix}$ and $n_2 = \begin{pmatrix} \alpha \\ -1 \\ 3 \end{pmatrix}$, and we need $n_1 . n_2 = 0$.

So, we require $1 \times \alpha + 2 \times (-1) + (-1) \times 3 = 0$. This gives $\alpha = 5$.

DON'T FORGET

When calculating the angle between vectors a and b, always start with $a.b$. If this is zero, then the angle between them is 90° without further calculation.

NOW TRY THIS

(1) (a) Use Gaussian elimination to solve the system of equations

$$2x - y + z = 8, \quad x + 2y + z = 3, \quad -x + 3y + 2z = 1.$$ **5**

   (b) Show that the line of intersection, L, of the planes $2x - y + z = 8$ and $x + 2y + z = 3$ has parametric equations

$$x = 5 - 3t, \quad y = -t, \quad z = 5t - 2.$$ **2**

   (c) Calculate the acute angle between the line L and the plane $-x + 3y + 2z = 1$. **4**

(2) (a) Find an equation for the plane π_1 which contains the points $A(1, 1, 0)$, $B(3, 1, -1)$ and $C(2, 0, -3)$. **4**

   (b) Given that π_2 is the plane with equation $x + 2y + z = 3$, calculate the size of the acute angle between the planes π_1 and π_2. **3**

(3) (a) Find an equation of the plane π_1 through the points $A(1, 1, 1)$, $B(2, -1, 1)$ and $C(0, 3, 3)$. **3**

   (b) The plane π_2 has equation $x + 3y - z = 2$.

   Given that the point $(0, a, b)$ lies on both the planes π_1 and π_2, find the values of a and b. Hence find an equation of the line of intersection of the planes π_1 and π_2. **4**

   (c) Find the size of the acute angle between the planes π_1 and π_2. **3**

www.intmath.com/Differential-equations/Predicting-AIDS.php

An equation for an unknown function y of x which involves derivatives of y is called a **differential equation**. The highest derivative in the equation determines its order.

The equation $\frac{dy}{dx} = x + y$ is a first-order equation.

The equation $\frac{d^2 y}{dx^2} + \frac{dy}{dx} - y = \sin x$ is a second-order equation.

FIRST-ORDER EQUATIONS

The easiest equations to deal with are called **separable**. In these equations, we can move all the y terms to one side of the equation and all the x terms to the other side.

When a differential equation is not separable, it is only possible to solve the equation in special cases, some of which will be considered in subsequent sections.

FIRST-ORDER SEPARABLE EQUATIONS

First-order separable equations are equations which can be put in the form:

$$f(y)\frac{dy}{dx} = g(x) \qquad \text{where } f \text{ and } g \text{ are given functions.}$$

Integrating both sides gives:

$$\int f(y)\,dy = \int g(x)\,dx.$$

The short-hand notation is acceptable in the examination.

Integrating both sides gives: $\qquad F(y) = G(x) + c$

where c is an arbitrary constant. This is the **general solution**.

The general solution of a first-order differential equation contains one arbitrary constant.

From this, we can usually express y as a function of x, although sometimes the solution has to be left in implicit form.

If we are given the value of y for a given value of x (this is called an **initial condition**), i.e. $y = y_0$ when $x = x_0$, we can obtain the value of the constant c.

DON'T FORGET

$f(y)dy = g(x)dx \Rightarrow$
$\int f(y)dy = \int g(x)dx$

DON'T FORGET

Remember the arbitrary constant 'c'.

Example 10.1

Obtain the general solution of the differential equation $\sec^2 y \frac{dy}{dx} = 2x$.

$$\int \sec^2 y \, dy = \int 2x \, dx$$

i.e. $\tan y = x^2 + c$, so: $\qquad y = \tan^{-1}(x^2 + c)$.

Example 10.2

Obtain the general solution of the differential equation $\frac{dy}{dx} = \frac{x}{y+1}$, and the particular solution for which $y = 1$ when $x = 0$.

Rearranging, we have: $\qquad (y+1)dy = xdx$, so: $\qquad \int (y+1)dy = \int xdx$

$$\Rightarrow \frac{(y+1)^2}{2} = \frac{x^2}{2} + c.$$

This can be written as:

$$(y+1)^2 = x^2 + k, \text{ where } k (= 2c) \text{ is the arbitrary constant.}$$

contd

FIRST-ORDER SEPARABLE EQUATIONS   contd

Solving for y gives: $y = -1 \pm \sqrt{x^2 + k}$.

There are two families of solutions: $y = -1 + \sqrt{x^2 + k}$ and $y = -1 - \sqrt{x^2 + k}$.

For the particular solution $y = 1$ when $x = 0$, we get $k = 4$.

We must take the $+$ sign, because the negative sign makes $y < 0$.

So, $y = -1 + \sqrt{x^2 + 4}$.

DON'T FORGET

Add c as soon as you integrate before any attempt to solve for y in terms of x.

Laws of growth and decay

An important application of separable first-order differential equations arises in simple models of growth and decay.

Example 10.3

A circular fungus grows by absorbing nutrients at its boundary. The radius of the fungus r is increasing at the rate $10 - r$ mm per day. At time $t = 0$, the fungus has radius 1 mm.

Obtain r as a function of t. Does the fungus grow without limit?

$$\frac{dr}{dt} = 10 - r, \text{ so: } \quad \frac{dr}{10 - r} = dt \quad \Rightarrow \quad -\ln|10 - r| = t + c.$$

When $t = 0$, $r = 1$, so: $c = -\ln 9 \quad \Rightarrow \quad -\ln|10 - r| = t - \ln 9$.

Hence: $\ln 9 - \ln|10 - r| = t \quad \Rightarrow \quad \ln\left|\frac{9}{10 - r}\right| = t$.

So: $\dfrac{9}{10 - r} = e^t$

and solving for r gives: $r = 10 - 9e^{-t}$.

From this, we can see that $r < 10$ for all t, and so the fungus will never have a radius greater than 10 mm.

If a population has unlimited resources to enable it to grow, the rate at which the population grows is proportional to the size of the population. So, if the size is N at time t, we have:

$$\frac{dN}{dt} = kN, \text{ where } k \text{ is a positive constant.}$$

We have $\dfrac{dN}{N} = k\,dt \quad \Rightarrow \quad \ln N = kt + c, \quad$ so $N = e^{kt + c} = e^{kt}\,e^c = Ae^{kt}$ (where $A = e^c$).

A is a positive arbitrary constant, so N increases exponentially.

The growth in world population has followed this type of exponential model approximately, over certain periods, although no single model of this type fits world population size over the past 200 years.

A model of population growth which has an 'overcrowding' factor is: $\dfrac{dN}{dt} = kN(N_0 - N)$

where k and N_0 are known positive constants. To solve this differential equation, we need partial fractions.

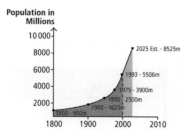

World Population Growth

Population in Millions

Example 10.4

Solve the differential equation $\dfrac{dN}{dt} = N(100 - N)$ given that $N(0) = 50$. What happens to N as $t \to +\infty$?

$$\frac{dN}{N(100 - N)} = dt \quad \Rightarrow \quad \int \frac{dN}{N(100 - N)} = t + c.$$

contd

DIFFERENTIAL EQUATIONS

Using partial fractions: $\dfrac{1}{N(100-N)} = \dfrac{1}{100}\left(\dfrac{1}{N} + \dfrac{1}{100-N}\right)$, so:

$$\int \frac{dN}{N(100-N)} = \frac{1}{100}\int\left(\frac{1}{N} + \frac{1}{100-N}\right)dN = \frac{1}{100}\Big[\ln N - \ln|100-N|\Big]$$

$$= \frac{1}{100}\ln\left|\frac{N}{100-N}\right|.$$

Hence: $\dfrac{1}{100}\ln\left|\dfrac{N}{100-N}\right| = t + C$, and $N = 50$ when $t = 0$ gives $C = 0$.

So: $\ln\left|\dfrac{N}{100-N}\right| = 100t \;\Rightarrow\; \dfrac{N}{100-N} = e^{100t}$ (we have $N < 100$).

Solving for N gives $N = e^{100t}(100-N)$, so: $N + Ne^{100t} = 100e^{100t}$.

Hence $N = \dfrac{100e^{100t}}{1+e^{100t}} = \dfrac{100}{1+e^{-100t}} \;\Rightarrow\; 100$ as $t \to +\infty$.

INTEGRATING FACTORS

This type of equation cannot be solved by the previous method: $\dfrac{dy}{dx} + a(x)y = b(x)$.

By multiplying by a suitable factor of x, the left-hand side can be made into an exact derivative, and then both sides of the equation can be integrated. This factor is called the **integrating factor** (IF).

> **DON'T FORGET**
>
> The integrating factor for $\dfrac{dy}{dx} + a(x)y = b(x)$ is $e^{\int a(x)dx}$.

Note that: $p(x)\dfrac{dy}{dx} + q(x)y = r(x)$

may be written as: $\dfrac{dy}{dx} + \dfrac{q(x)}{p(x)}y = \dfrac{r(x)}{p(x)}$

so: $a(x) = \dfrac{q(x)}{p(x)}$ and $b(x) = \dfrac{r(x)}{p(x)}$.

The differential equation above, in which the coefficient of $\dfrac{dy}{dx}$ is 1, is called the **standard form**. This is called a **linear differential equation** because there are no powers of y or products of y with $\dfrac{dy}{dx}$.

Steps for solving equations using the integrating factor

> **DON'T FORGET**
>
> The constant of integration before dividing by the IF

1 Write in standard form
$\dfrac{dy}{dx} + a(x)y = b(x)$

2 Obtain IF (the integrating factor) where $IF = e^{\int a(x)dx}$

3 Multiply the standard form by IF to get:
$IF\dfrac{dy}{dx} + IFa(x)y = IFb(x)$

4 $\dfrac{d}{dx}(IF.y) = IF.b(x)$

5 $IF.y = \int IF.b(x)\,dx$

6 $y = \dfrac{\int IF.b(x)\,dx}{IF}$

Example 10.5

> **DON'T FORGET**
>
> After multiplying by the integrating factor, the left-hand side of the equation will be the derivative of a product.

Obtain the general solution of the equation $\dfrac{dy}{dx} + \dfrac{2}{x}y = \dfrac{\sin x}{x^2}$ $(x > 0)$.

$\int \dfrac{2}{x}dx = 2\ln x = \ln(x^2)$, so the integrating factor is: $e^{\ln(x^2)} = x^2$.

So: $x^2\dfrac{dy}{dx} + 2xy = \sin x$, i.e. $\dfrac{d}{dx}(x^2y) = \sin x \;\Rightarrow\; x^2y = -\cos x + C$,

and so the general solution is: $y = -\dfrac{\cos x}{x^2} + \dfrac{C}{x^2}$.

contd

INTEGRATING FACTORS   contd

Example 10.6

Obtain the solution of $\sin x \dfrac{dy}{dx} - 2y\cos x = 3\sin^3 x$ for which $y = 0$ when $x = \dfrac{\pi}{2}$.

We first need to divide through by $\sin x$ to put the equation in the correct form:

$\dfrac{dy}{dx} - \dfrac{2\cos x}{\sin x} y = 3\sin^2 x$, so the integrating factor is:

$$\exp\left(\int \dfrac{-2\cos x}{\sin x}\,dx\right) = \exp(-2\ln\sin x) = \exp(\ln(\sin x)^{-2}) = \dfrac{1}{\sin^2 x}.$$

Don't forget the minus sign.

Multiplying by the integrating factor gives: $\quad \dfrac{1}{\sin^2 x}\dfrac{dy}{dx} - \dfrac{2\cos x}{\sin^3 x} y = 3,$

so: $\quad \dfrac{d}{dx}\left(\dfrac{y}{\sin^2 x}\right) = 3 \implies \dfrac{y}{\sin^2 x} = 3x + C$

and the general solution is: $\quad y = (3x + C)\sin^2 x.$

As $y = 0$ when $x = \dfrac{\pi}{2}$, $C = -\dfrac{3\pi}{2}$, giving the solution $y = \left(3x - \dfrac{3\pi}{2}\right)\sin^2 x.$

DON'T FORGET

The constant C must be introduced when you integrate $b(x)e^{A(x)}$.

NOW TRY THIS

(1) Solve the differential equation

$$x\dfrac{dy}{dx} - 2y = x^3 e^{2x}$$

given that $y(1) = 0$. Express your answer in the form $y = f(x)$.   6

(2) A garden centre advertises young plants to be used as hedging.

After planting, the growth G metres (i.e. the increase in height) after t years is modelled by the differential equation

$$\dfrac{dG}{dt} = \dfrac{25k - G}{25}$$

where k is a constant and $G = 0$ when $t = 0$.

(a) Express G in terms of t and k.   4

(b) Given that a plant grows 0·6 metres by the end of 5 years, find the value of k correct to 3 decimal places.   2

(c) On the plant labels, it states that the expected growth after 10 years is approximately 1 metre. Is this claim justified?   2

(d) Given that the initial height of the plants was 0·3 m, what is the likely long-term height of the plants?   2

(3) (a) A mathematical biologist believes that the differential equation

$$x\dfrac{dy}{dx} - 3y = x^4$$

models a process. Find the general solution of the differential equation.   5

Given that $y = 2$ when $x = 1$, find the particular solution, expressing y in terms of x.   2

(b) The biologist subsequently decides that a better model is given by the equation

$$y\dfrac{dy}{dx} - 3x = x^4.$$

Given that $y = 2$ when $x = 1$, obtain y in terms of x.   4

SECOND-ORDER EQUATIONS WITH CONSTANT COEFFICIENTS

HOMOGENEOUS SECOND-ORDER DIFFERENTIAL EQUATIONS

The equation $a\dfrac{d^2y}{dx^2} + b\dfrac{dy}{dx} + cy = 0$.

This is called a **homogeneous equation** because the right-hand side is zero.

The general solution of this equation depends on the quadratic equation $am^2 + bm + c = 0$. This is called the **auxiliary equation**. There are three separate cases to consider, depending on the nature of the roots. These three cases are:

- real unequal roots
- equal roots
- complex roots.

General solutions for these three cases are shown below. Note that A and B are arbitrary constants.

> **DON'T FORGET**
>
> The general solution of second-order differential equations requires two arbitrary constants.

1 Real unequal roots α, β. The general solution is: $y = Ae^{\alpha x} + Be^{\beta x}$

2 Equal roots α, α. The general solution is: $y = Ae^{\alpha x} + Bxe^{\alpha x}$ (or $y = (A + Bx)e^{\alpha x}$)

3 Complex roots $p \pm qi$. The general solution is: $y = e^{px}(A\cos qx + B\sin qx)$

Example 10.7

Obtain the general solution of the equation $\dfrac{d^2y}{dx^2} - 4\dfrac{dy}{dx} - 5y = 0$.

The auxiliary equation is:

$$m^2 - 4m - 5 = 0$$
$$\Rightarrow (m - 1)(m + 5) = 0.$$

The roots are 1 and −5, so the general solution is:

$$y = Ae^x + Be^{-5x}.$$

To evaluate two arbitrary constants, we need two pieces of information which come in the form of initial conditions, such as where the values of y and $\dfrac{dy}{dx}$ are given for a given value of x.

Example 10.8

Obtain the solution of $\dfrac{d^2y}{dx^2} - 4\dfrac{dy}{dx} + 4y = 0$ for which $y = 2$ and $\dfrac{dy}{dx} = 1$ when $x = 0$.

The auxiliary equation is:

$$m^2 - 4m + 4 = 0$$
$$(m - 2)^2 = 0.$$

This gives roots = 2, 2.

The general solution is $y = (A + Bx)e^{2x}$

$$\frac{dy}{dx} = Be^{2x} + 2(A + Bx)e^{2x}.$$

Setting $y = 2$ when $x = 0$ gives $A = 2$.

$\dfrac{dy}{dx} = 1$ when $x = 0$ gives: $B + 2A = 1$, so $B = -3$.

The solution is: $y = (2 - 3x)e^{2x}$.

contd

HOMOGENEOUS SECOND-ORDER DIFFERENTIAL EQUATIONS   contd

Example 10.9

Obtain the solution of $\frac{d^2y}{dx^2} + 8\frac{dy}{dx} + 20y = 0$ given that $y = 1$ when $x = 0$ and $y = 2$ when $x = \frac{\pi}{4}$.

The auxiliary equation is:

$$m^2 + 8m + 20 = 0.$$

By completing the square, $(m + 4)^2 = -4$, so:

$$m = -4 \pm 2i.$$

Alternatively, by using the quadratic formula:

$$m = \frac{-8 \pm \sqrt{8^2 - 4.1.20}}{2} = \frac{-8 \pm 4i}{2} = -4 \pm 2i.$$

Hence the general solution is:

$$y = e^{-4x}(A\cos 2x + B\sin 2x).$$

Setting $y = 1$ when $x = 0$ gives $A = 1$, and $y = 2$ when $x = \frac{\pi}{4}$ gives $B = 2e^{\pi}$.

The solution is:

$$y = e^{-4x}(\cos 2x + 2e^{\pi}\sin 2x).$$

NON-HOMOGENEOUS SECOND-ORDER DIFFERENTIAL EQUATIONS

To solve $a\frac{d^2y}{dx^2} + b\frac{dy}{dx} + cy = f(x),$

first solve the homogeneous equation $a\frac{d^2y}{dx^2} + b\frac{dy}{dx} + cy = 0.$

This gives the **complementary function (CF)**.

We then find a **particular integral (PI)** (methods given below) which leads to the **general solution** of the non-homogeneous equation.

General solution = complementary function + particular integral.

DON'T FORGET

General solution = CF + PI

Example 10.10

Verify that $x + 1$ is a particular integral of the equation:

$$\frac{d^2y}{dx^2} - 3\frac{dy}{dx} + 2y = 2x - 1.$$

and hence obtain the general solution of this equation.

If $y = x + 1$, $\frac{dy}{dx} = 1$ and $\frac{d^2y}{dx^2} = 0,$ so:

$$\frac{d^2y}{dx^2} - 3\frac{dy}{dx} + 2y = 0 - 3(1) + 2(x + 1) = 2x - 1.$$

The auxiliary equation for $\frac{d^2y}{dx^2} - 3\frac{dy}{dx} + 2y = 0$ is:

$$m^2 - 3m + 2 = 0,$$ which has roots 1 and 2, and general solution:

$$y = Ae^x + Be^{2x}.$$

Hence the general solution of $\frac{d^2y}{dx^2} - 3\frac{dy}{dx} + 2y = 2x - 1$ is:

$$y = x + 1 + Ae^x + Be^{2x}.$$

contd

DIFFERENTIAL EQUATIONS

Methods for finding a particular integral

For certain types of function f, it is possible to guess the general form of the particular integral and then evaluate any unknown parameters. This is not the same as evaluating arbitrary constants in a general solution. The table below should be remembered.

Forms of $f(x)$	Forms of particular integral
Polynomials	
$3x + 4$	$ax + b$
$6x^2 + 2$	$ax^2 + bx + c$
$10x^2 + 3x$	$ax^2 + bx + c$
Trigonometric functions	
$4 \cos x$	$a\cos x + b\sin x$
$3 \sin 2x$	$a\cos 2x + b\sin 2x$
Exponential functions	
$2e^{3x}$	ae^{3x}

Example 10.11

Obtain the general solution of the equation $\dfrac{d^2y}{dx^2} + 3\dfrac{dy}{dx} - 4y = 4e^{-x}$, and hence obtain the particular solution satisfying $y = 2$ and $\dfrac{dy}{dx} = 3$ when $x = 0$.

The auxiliary equation for the homogeneous equation is:

$m^2 + 3m - 4 = 0$, which has roots 1 and -4, so the complementary function is:

$y = Ae^x + Be^{-4x}$.

Try as a particular integral $y = ae^{-x}$:

$$\frac{dy}{dx} = -ae^{-x}, \qquad \frac{d^2y}{dx^2} = ae^{-x}.$$

Substituting into the differential equation:

$$ae^{-x} - 3ae^{-x} + 4ae^{-x} = 4e^{-x} \quad \Rightarrow \quad 2ae^{-x} = 4e^{-x},$$

so: $a = 2$.

The general solution of $\dfrac{d^2y}{dx^2} + 3\dfrac{dy}{dx} - 4y = 4e^{-x}$ is:

$y =$ CF + PI   so: $y = Ae^x + Be^{-4x} + 2e^{-x}$.

If $y = 2$ when $x = 0$, then $A + B = 0$.

Also:

$$\frac{dy}{dx} = Ae^x - 4Be^{-4x} - 2e^{-x}, \text{ so:}$$

$\dfrac{dy}{dx} = 3$ when $x = 0$ gives $A - 4B = 5$.

Solving for A and B gives: $A = 1, B = -1$.

The solution satisfying $y = 2$ and $\dfrac{dy}{dx} = 3$ when $x = 0$ is:

$y = e^x - e^{-4x} + 2e^{-x}$.

DON'T FORGET

Any initial conditions must be applied to the general solution of the full equation (CF + PI).

contd

NON-HOMOGENEOUS SECOND-ORDER DIFFERENTIAL EQUATIONS   contd

Example 10.12

Obtain the general solution of the equation:

$$\frac{d^2y}{dx^2} + 6\frac{dy}{dx} + 25y = 2\sin x.$$

The auxiliary equation for the homogeneous equation is:

$$m^2 + 6m + 25 = 0.$$

Completing the square (or using the quadratic formula) gives:

$(m+3)^2 = -16 \implies m = -3 \pm 4i,$ so the general solution of the homogeneous equation (i.e. the complementary function) is: $y = e^{-3x}(A\cos 4x + B\sin 4x).$

For a particular integral of the full equation, let $y = a\cos x + b\sin x$:

$$\frac{dy}{dx} = -a\sin x + b\cos x \quad \text{and} \quad \frac{d^2y}{dx^2} = -a\cos x - b\sin x.$$

Substituting into the full equation gives:

$$-a\cos x - b\sin x + 6(b\cos x - a\sin x) + 25(a\cos x + b\sin x) = 2\sin x.$$

This is an identity, so, after collecting like terms on the left-hand side, we equate the coefficients of $\cos x$ and $\sin x$ on both sides:

$$\cos x: \quad -a + 6b + 25a = 0$$

$$\sin x: \quad -b - 6a + 25b = 2.$$

Solving the two equations $24a + 6b = 0$ and $-6a + 24b = 2$ simultaneously gives:

$$a = -\frac{1}{51}, \quad b = \frac{4}{51}.$$

So, the general solution of $\frac{d^2y}{dx^2} + 6\frac{dy}{dx} + 25y = 2\sin x$ is:

$$y = e^{-3x}(A\cos 4x + B\sin 4x) + \frac{4}{51}\sin x - \frac{1}{51}\cos x.$$

NOW TRY THIS

(1) Solve the differential equation

$$\frac{d^2y}{dx^2} + 6\frac{dy}{dx} + 9y = 0$$
4

given that $y = 2$ and $\frac{dy}{dx} = -3$ when $x = 0$.
3

(2) Obtain the general solution of the differential equation

$$\frac{d^2y}{dx^2} + 4\frac{dy}{dx} - 5y = 7e^{2x} + 10.$$
7

Obtain the particular solution for which $y(0) = 1$ and $y'(0) = 10$.
3

(3) Obtain the general solution of the differential equation

$$y'' + 4y' + 8y = 8x^2 + 16x + 6 \quad (' = \frac{d}{dx}).$$
6

(4) Solve the differential equation

$$\frac{d^2y}{dx^2} + 2\frac{dy}{dx} + 5y = 0$$

given that $y = 1$ when $x = 0$ and $y = 2$ when $x = \frac{\pi}{4}$.
6

NUMBERS AND DIRECT PROOF

Numbers are classified according to type, and you need to be familiar with natural numbers, integers, primes, rational and irrational numbers, real numbers and complex numbers.

NATURAL NUMBERS

Natural numbers are the counting numbers, 1, 2, 3, … . The set of **natural numbers** is often denoted by $\mathbb{N}$, so the notation to show that 2 is a natural number is: $2 \in \mathbb{N}$ (remembering that $\in$ means 'is a member of').

INTEGERS

When we add 0 and the negative numbers -1, -2, … to the natural numbers, we get the set of **integers**, usually denoted by $\mathbb{Z}$. $\mathbb{N}$ is a subset of $\mathbb{Z}$. So, $-2 \in \mathbb{Z}$.

An important subset of $\mathbb{Z}$, and hence also of $\mathbb{R}$, is the set of **prime numbers**, $\mathbb{P}$. These are the positive integers which cannot be factorised. The first six are 2, 3, 5, 7, 11, 13.

 http://mathworld.wolfram.com/PrimeNumber.html

The fundamental theorem of arithmetic

The fundamental theorem of arithmetic states that every integer greater than 1 can be written as a product of primes in one, and only one, way.

 http://mathworld.wolfram.com/FundamentalTheoremofArithmetic.html
www.mathacademy.com/pr/prime/articles/fta/index.asp

Example 11.1

Factorise 16 500.

As the number is clearly even, start by dividing by 2 and continue for as long as possible:

$16\,500 \div 2 \div 2 = 4125$, which is odd, so we now try dividing by 3:

$4125 \div 3 = 1375$, but this is not divisible by 3, so next try 5:

$1375 \div 5 \div 5 = 11$, and as 11 is prime, we stop.

$16\,500 = 2^2 \times 3 \times 5^3 \times 11$.

RATIONAL AND IRRATIONAL NUMBERS

Rational numbers are of the form $\frac{m}{n}$ ($n \neq 0$), where m and n are integers.

This set is often denoted by $\mathbb{Q}$. $\mathbb{Z}$ is a subset of $\mathbb{Q}$.

Irrational numbers cannot be expressed in the form $\frac{m}{n}$. An example of an irrational number is $\sqrt{2}$.

 http://mathworld.wolfram.com/IrrationalNumber.html

REAL NUMBERS

Real numbers are either rational or irrational. The set of all real numbers is usually denoted by $\mathbb{R}$. $\mathbb{Q}$ is a subset of $\mathbb{R}$.

COMPLEX NUMBERS

Complex numbers are of the form $x + iy$, where x and y are real numbers, and $i = \sqrt{-1}$.
The set of complex numbers is usually denoted by $\mathbb{C}$. $\mathbb{R}$ is a subset of $\mathbb{C}$.

DIRECT PROOF

The simplest and most straightforward type of proof is direct proof.

Example 11.2

This is a simple, *direct proof* that the sum of two even integers is itself an even number.

Consider two even integers x and y.

Since they are even, they can be written as $x = 2a$ and $y = 2b$ respectively for integers a and b.

Then the sum $x + y = 2a + 2b = 2(a + b)$.

From this, it is clear that $x + y$ has 2 as a factor and therefore is even, so the sum of any two even integers is even.

Example 11.3

The sum of any two rational numbers is rational.

This proof follows directly from the definition of what it means for a number to be rational. Given that r and s are rational numbers, show that $r + s$ is rational.

Proof: Since r and s are rational, we can write $r = \frac{p}{q}$ and $s = \frac{m}{n}$ for some integers p, q, m and n.

Then $r + s = \frac{p}{q} + \frac{m}{n}$

$\qquad = \frac{pn + qm}{qn}$.

Since p, q, m and n are integers, $pn + qm$ and qn are also integers (the sum or product of integers is an integer). Thus, by the calculation above, $r + s$ is the quotient of two integers, and is therefore a rational number.

Example 11.4

Here's an example of a proof that is really just a calculation.

Given the trigonometric identity $\sin(x + y) = \sin x \cos y + \cos x \sin y$, prove the identity: $\sin(2x) = 2 \sin x \cos x$.

Proof: $\sin(2x) = \sin(x + x)$

$\qquad\qquad = \sin x \cos x + \cos x \sin x \qquad$ (by the sum identity)

$\qquad\qquad = 2 \sin x \cos x$.

Note that this proof is merely a string of equalities connecting $\sin(2x)$ to $2 \sin x \cos x$. Many proofs are like this when proving an identity.

⚙ NOW TRY THIS

(1) Prove that $\sec^2 x = 1 + \tan^2 x$. 　　　　　　　　　　　　　　　　　　　3

PROOF BY CONTRADICTION

Suppose we have two statements A and B, and we want to prove that if A is true then B is true (we write this as $A \Rightarrow B$, and say 'if A then B'). One way of doing this is to assume that A is true but that B is false, and get a contradiction. This is the method of **proof by contradiction**.

Example 11.5

Consider these two statements:

A n^2 is even

B n is even ($n \in \mathbb{Z}$).

Prove that $A \Rightarrow B$ by contradiction.

Assume that A is true but B is false. Then n is odd and $n = 2m + 1$ where $m \in \mathbb{Z}$.

But: $(2m + 1)^2 = 4m^2 + 4m + 1$

$$= 2(2m^2 + 2m) + 1$$

$$= 2r + 1$$

where $r \in \mathbb{Z}$.

This shows that $(2m + 1)^2 = n^2$ is odd, which contradicts statement A.

Hence n must be even.

> **DON'T FORGET**
>
> An even integer can be written as $2m$ and an odd integer as $2m + 1$, where m is an integer.

Example 11.6

A common question in assessments is to prove that, for some non-square integer a, $\sqrt{a}$ is irrational.

The first method below holds for all such a.

The second method is shorter for the special case where a is even.

Method 1

Prove by contradiction that $\sqrt{7}$ is irrational.

Since $\sqrt{7}$ must be either rational or irrational, assume that it is rational.

Then $\sqrt{7} = \frac{m}{n}$ for integers m and n, which have no common factors because we can cancel any common factor first, $[(m, n) = 1]$.

$$\sqrt{7} = \frac{m}{n}$$

$$7 = \frac{m^2}{n^2}$$

$$7n^2 = m^2$$

$\therefore 7 | m$ by the fundamental theorem of arithmetic

$\therefore 7^2 | m^2$

$\therefore 7^2 | 7n^2$

$\therefore 7 | n^2$

$\therefore 7 | n$ by the fundamental theorem of arithmetic

$\therefore 7 | m$ and $7 | n$, i.e. m and n have a common factor.

This contradicts $(m, n) = 1$.

This in turn contradicts our original assumption that $\sqrt{7}$ is rational, and hence $\sqrt{7}$ must be irrational.

Method 2

Prove by contradiction that $\sqrt{6}$ is irrational.

Since $\sqrt{6}$ must be either rational or irrational, assume that it is rational.

Then $\sqrt{6} = \frac{m}{n}$ for integers m and n which have no common factors, because we can cancel any common factor first: $[(m, n) = 1]$.

So:

$$6 = \frac{m^2}{n^2} \Rightarrow m^2 = 6n^2$$

$$\Rightarrow m^2 = 2 \times 3n^2$$

and hence m^2 is even.

By Example 11.5, it follows that m is even, $m = 2r \quad (r \in \mathbb{Z})$.

So, $4r^2 = m^2 = 6n^2 \quad \Rightarrow \quad 3n^2 = 2r^2$, and hence $3n^2$ is even.

But 3 is odd, so n^2 must be even, and hence n must be even.

This means that both m and n are even, which is a contradiction because we started with m and n having no common factor. So, the assumption that $\sqrt{6}$ is rational has led to a contradiction, and hence $\sqrt{6}$ must be irrational.

http://nrich.maths.org/4717
www.math.jmu.edu/~taal/235_2000post/235contradiction.pdf
http://mathworld.wolfram.com/ProofbyContradiction.html
http://delphiforfun.org/programs/Math_Topics/proof_by_contradiction.htm

NOW TRY THIS

(1) Prove by contradiction that, if x is an irrational number, $\frac{1}{x}$ is irrational.   **4**

FURTHER PROOFS

COUNTEREXAMPLES

To prove a mathematical statement, it is not enough to list numerical examples which support it, however many you give – because just one example is enough to disprove a statement. Such an example is called a **counterexample**. This is a useful method to disprove a statement which claims to be true 'for all possible values of n, $\forall n$'.

Example 11.7

Prove or disprove the statement: 'If a function f with domain $\mathbb{R}$ satisfies $f''(0) = 0$, then f has a point of inflexion at $x = 0$'.

The statement is false. A counterexample is the function $f(x) = x^4$ (because f'' doesn't change sign at $x = 0$, the curve remains concave up).

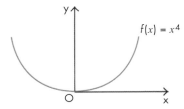

$f''(x) = 12x^2 \geqslant 0$, so always concave up; $f''(0) = 0$, but $x = 0$ is not a point of inflexion.

Example 11.8

Consider the statement: 'If the product of two integers m and n is even, both integers are even'. Either prove this result or give a counterexample.

The statement isn't true; one counterexample is $m = 2$, $n = 3$.

You can try out your knowledge at

 http://webspace.ship.edu/deensley/DiscreteMath/flash/ch2/sec2_1/
counterexamples/control21.html

NECESSARY AND SUFFICIENT

Consider two statements A and B. If $A \Rightarrow B$, we say that A is sufficient for B and that B is necessary for A (because if B isn't true, A can't be either).

Sometimes we have both $A \Rightarrow B$ and $B \Rightarrow A$. In this case, we write $A \Leftrightarrow B$ and say that A is necessary and sufficient for B.

Example 11.9

Consider these statements A and B about integers m and n:

　　A　　m and n are odd

　　B　　mn is odd.

Decide whether A is (a) sufficient, (b) necessary, (c) necessary and sufficient for B.

(a) does $A \Rightarrow B$?

　　Let $m = 2p + 1$, $n = 2q + 1$ (p, q integers).

　　($2p + 1$ and $2p - 1$ would not be general enough; we need distinct p, q).

　　Then $mn = (2p + 1)(2q + 1) = 4pq + 2p + 2q + 1 = 2(2pq + p + q) + 1$, which is odd.

　　Hence A is sufficient for B.

contd

NECESSARY AND SUFFICIENT contd

(b) does $B \Rightarrow A$? We prove this is true by contradiction. Assume that mn is odd, and suppose at least one of m, n is even. If $m = 2p$ then $mn = 2pn$, which is even, contradicting the assumption that mn is odd. Hence $B \Rightarrow A$, and A is necessary for B.

(c) Because both $A \Rightarrow B$ and $B \Rightarrow A$, we have $A \Leftrightarrow B$.

Hence A is necessary and sufficient for B.

> **DON'T FORGET**
>
> A sufficient for B means $A \Rightarrow B$.
>
> A necessary for B means $B \Rightarrow A$.

IF AND ONLY IF

Another way of formulating 'necessary and sufficient' statements is to use 'if and only if'.

Example 11.10

Prove that a polynomial $p(x) = a + bx + cx^2$ is an even function if and only if $b = 0$.

There are two parts to an 'if and only if' proof:

'if' part:

If $b = 0$, then $p(x) = a + cx^2$,

so $p(-x) = a + c(-x)^2 = a + cx^2 = p(x)$

$p(-x) = p(x) \implies p$ is an even function.

'only if' part:

Let the function p be even, so that $p(x) = p(-x)$ for all x.

Then $a + bx + cx^2 = a - bx + cx^2 \implies bx = -bx$ for all x

which gives $b = -b$, and so $b = 0$.

> **DON'T FORGET**
>
> These statements are all equivalent:
>
> - $A \Leftrightarrow B$
> - A is necessary and sufficient for B
> - A is true if and only if B is true

> **DON'T FORGET**
>
> An 'if and only if' proof consists of two parts.

NOW TRY THIS

(1) For each of the following statements, decide if it is true or false. Justify your conclusions.

    A The cube of any even integer p plus the square of any odd integer q is odd.

    B For all natural numbers m, if m^2 is divisible by 9, then m is divisible by 9.   **5**

(2) Consider the number abc in base 10 (i.e. the integer $100a + 10b + c$).
Prove that abc is divisible by 9 if, and only if, $a + b + c$ is divisible by 9.   **5**

(3) Given that $p(n) = n^2 + n$, where n is a positive integer, consider the statements:

    A $p(n)$ is always even

    B $p(n)$ is always a multiple of 3.

    For each statement, prove it if it is true or, otherwise, disprove it.   **4**

PROOF BY INDUCTION

Proof by induction is often used to prove results involving sums of terms – **Σ-type proofs**.

Example 11.11

Prove by induction that $4 + 7 + 10 + \ldots + (3n + 1) = \frac{1}{2}n(3n + 5)$ for all integers $n \geqslant 1$.

This can also be written using Σ notation as:

$$4 + 7 + 10 + \ldots + (3n + 1) = \sum_{r=1}^{n}(3r + 1).$$

Step 1: Check that the formula holds for $n = 1$:

LHS $= 4$ RHS $= \frac{1}{2}(1)(3 + 5) = 4$

$\therefore$ true for $n = 1$.

Step 2: Assume that the formula has been proved for some integer k (we know that $k = 1$ works), so that $4 + 7 + 10 + \ldots + (3k + 1) = \frac{1}{2}k(3k + 5)$.

We now need to show that, given this result, the result with k replaced by $k + 1$ holds as well, i.e.:

$$4 + 7 + \ldots + (3k + 1) + (3(k + 1) + 1) = \frac{1}{2}(k + 1)(3(k + 1) + 5)$$

$$4 + 7 + \ldots + (3k + 1) + (3k + 4) = \frac{1}{2}(k + 1)(3k + 8).$$

Step 3: We now have to do some algebra.

Because we are assuming the result for k, namely:

$$4 + 7 + 10 + \ldots + (3k + 1) = \frac{1}{2}k(3k + 5)$$

it follows that:

$$4 + 7 + \ldots + (3k + 1) + (3k + 4) = \frac{1}{2}k(3k + 5) + (3k + 4).$$

We now have to show that the right-hand side of the equation is indeed

$\frac{1}{2}(k + 1)(3k + 8)$.

Start with this RHS and take out a common factor:

$$\frac{1}{2}k(3k + 5) + (3k + 4)$$
$$= \frac{1}{2}[k(3k + 5) + (6k + 8)]$$
$$= \frac{1}{2}(3k^2 + 11k + 8)$$
$$= \frac{1}{2}(3k + 8)(k + 1)$$

$\therefore$ if true for $n = k$, then true for $n = k + 1$.

Step 4: We have shown that the result holds for $n = 1$, and that when the result holds for $n = k$, it holds for $n = k + 1$. So, true for $n = 1$ $\implies$ true for $n = 2$, true for $n = 2$ $\implies$ true for $n = 3$, and so on. Hence, by induction, the result is true for all n.

The induction method can also be used for proofs involving inequalities or divisibility statements.

Example 11.12

Prove by induction that $n^2 > 3n + 1$ for all integers $n \geq 4$.

Step 1: True for $n = 4$ because $16 > 13$. Note that the result isn't true for $n = 1, 2, 3$.

Step 2: Assume that $k^2 > 3k + 1$ for some integer k. We need to deduce from this that $(k+1)^2 > 3(k+1) + 1 = 3k + 4$.

Step 3: This is where we do the algebra:

$(k+1)^2 = k^2 + 2k + 1 > (3k+1) + 2k + 1$ because we are assuming $k^2 > 3k + 1$.

So: $(k+1)^2 > (3k+1) + 2k + 1 = (3k+4) + (2k-2)$

and because $k > 1$ (actually $k \geq 4$): $(3k+4) + (2k-2) > 3k + 4$.

Hence we have shown that $k^2 > 3k + 1$ implies that $(k+1)^2 > 3(k+1) + 1$.

Step 4: The result holds for $n = 4$, hence it holds for $n = 5$, and so on. Thus, by induction, the result holds for all $n \geq 4$.

Example 11.13

Prove by induction that $6^n + 4$ is divisible by 10 for all integers $n \geq 1$.

Step 1: True for $n = 1$, because $6 + 4 = 10$. $10 | 10$

Step 2: Assume for some k that $6^k + 4$ is divisible by 10, so that:

$6^k + 4 = 10m$ (or $6^k = 10m - 4$)   for some integer m.

Step 3: Consider $6^{k+1} + 4$: $6^{k+1} + 4 = 6 \cdot 6^k + 4$

$$= 6(10m - 4) + 4$$
$$= 60m - 20$$
$$= 10(6m - 2).$$

So, $6^k + 4$ divisible by 10 $\Rightarrow$ $6^{k+1} + 4$ divisible by 10.

Step 4: The result holds for $n = 1$, hence it holds for $n = 2$, hence for $n = 3$ etc.

So, by induction, $6^n + 4$ is divisible by 10 for all integers $n \geq 1$.

NOW TRY THIS

(1) Prove by induction that $\sum_{r=1}^{n} (6r^2 + 4r) = n(n+1)(2n+3)$.   **5**

(2) Prove by induction on n that, for $x > 0$,

$(1+x)^n \geq 1 + nx + \frac{1}{2}n(n-1)x^2$

for all positive integers n.   **5**

(3) Given that A and B are non-singular square matrices of same size, express the inverse of AB in terms of A^{-1} and B^{-1}.

Prove by induction on n that the inverse of A^n is $(A^{-1})^n$ for all integers $n \geq 1$.   **5**

(4) Prove by induction that $\frac{d^n}{dx^n}(xe^x) = (x+n)e^x$ for all integers $n \geq 1$.   **5**

EUCLIDEAN ALGORITHM

For any positive integers a and b ($b \neq 0$), there exist unique integers q and r, where $0 \leqslant r < b$, such that: $\quad a = bq + r$.

This is the **division algorithm**.

Example 11.14

If $a = 1127$ and $b = 17$, find q and r ($0 \leqslant r < 17$) such that $1127 = 17q + r$.

$\dfrac{1127}{17} = 66 \cdot 29$, so consider $1127 - 66 \times 17$.

This gives $1127 = 66 \times 17 + 5$.

The **greatest common divisor** (gcd) of two integers a and b is the largest integer that divides both a and b exactly.

When the gcd of a and b is 1, we say that they are **co-prime**.

The notation (a, b) for the gcd of a and b is often used.

Euclid's algorithm states that: $\quad$ if $a = bq + r$ where $0 \leqslant r < b$, then $(a, b) = (b, r)$.

Example 11.15

Obtain the gcd of 1800 and 210.

$$
\begin{aligned}
1800 &= 210 \times 8 + 120 & &\text{so } (1800, 210) = (210, 120) \\
210 &= 120 \times 1 + 90 & &\text{so } (210, 120) = (120, 90) \\
120 &= 90 \times 1 + 30 & &\text{so } (120, 90) = (90, 30) \\
90 &= 30 \times 3 + 0 & &\text{so } (90, 30) = (30, 0) = 30.
\end{aligned}
$$

Hence the gcd of 1800 and 210 is 30. The statements in red do not need to be included in a solution to an examination question, but you may find them helpful.

Example 11.16

Find integers s and t such that $1800s + 210t = 30$.
From Example 11.15, we have:

$$30 = 120 - 90 \times 1$$
$$30 = 120 - (210 - 120 \times 1)$$
$$30 = 2 \times 120 - 210$$
$$30 = 2 \times (1800 - 8 \times 210) - 210$$
$$30 = 2 \times 1800 - 17 \times 210$$
$$\therefore s = 2, t = -17.$$

The working for the two examples above may be combined as follows:

$1800 = 210 \times 8 + 120$	$\therefore 120 = 1800 - 210 \times 8$	[1]
$210 = 120 \times 1 + 90$	$\therefore 90 = 210 - 120$	[2]
$120 = 90 \times 1 + 30$	$\therefore 30 = 120 - 90$	[3]
$[2] \rightarrow [3]$	$30 = 120 - (210 - 120)$	
	$30 = 2 \times 120 - 210$	[4]
$[1] \rightarrow [4]$	$30 = 2(1800 - 210 \times 8) - 210$	
	$30 = 1800 \times 2 - 210 \times 17$	
$\therefore s = 2, t = -17.$		

Example 11.17

Find integers a and b such that $7425a + 2744b = 1$.

We start by finding the gcd of 7425 and 2744.

$$7425 = 2744 \times 2 + 1937 \qquad \text{so } (7425, 2744) = (2744, 1937)$$
$$2744 = 1937 \times 1 + 807 \qquad \text{so } (2744, 1937) = (1937, 807)$$
$$1937 = 807 \times 2 + 323 \qquad \text{so } (1937, 807) = (807, 323)$$
$$807 = 323 \times 2 + 161 \qquad \text{so } (807, 323) = (323, 161)$$
$$323 = 161 \times 2 + 1 \qquad \text{so } (323, 161) = (161, 1) = 1.$$

This shows that the gcd of 7425 and 2744 is 1. As before, the statements in red can be omitted.

Working backwards, we have:

$$1 = 323 - 161 \times 2$$
$$= 323 - 2 \times (807 - 323 \times 2)$$
$$= 5 \times 323 - 2 \times 807$$
$$= 5 \times (1937 - 807 \times 2) - 2 \times 807$$
$$= 5 \times 1937 - 12 \times (2744 - 1937)$$
$$= 17 \times 1937 - 12 \times 2744$$
$$= 17 \times (7425 - 2744 \times 2) - 12 \times 2744$$
$$= 17 \times 7425 - 46 \times 2744.$$

So, $a = 17$ and $b = -46$.

CHANGING NUMBER BASES

The division algorithm can be used to change a number represented in one base to its representation in another base. A number m in base a can be written as m_a.

Example 11.18

Express 315_6 as a number in base 3.

First, convert the number to base 10:

$$315_6 = \begin{array}{c|c|c} 36 & 6 & u \\ \hline 3 & 1 & 5 \end{array}$$

$$315_6 = 3 \times 6^2 + 1 \times 6 + 5 \times 6^0 = 119_{10}.$$

We now use repeated division by 3:

$$119 \div 3 = 39 \text{ remainder } 2$$
$$39 \div 3 = 13 \text{ remainder } 0$$
$$13 \div 3 = 4 \text{ remainder } 1$$
$$4 \div 3 = 1 \text{ remainder } 1$$
$$1 \div 3 = 0 \text{ remainder } 1$$
$$119_{10} = 11\,102_3$$

So, $315_6 = 11\,102_3$.

NOW TRY THIS

(1) Use the Euclidean algorithm to obtain the greatest common divisor of 29 400 and 6860, expressing it in the form $29\,400a + 6860b$, where a and b are integers.　　**4**

(2) Use the Euclidean algorithm to express 426_7 as a number in base 5.　　**4**

(3) Use the Euclidean algorithm to show that $(231, 17) = 1$, where (a, b) denotes the highest common factor of a and b.

Hence find integers x and y such that $231x + 17y = 1$.　　**4**

ADVANCED HIGHER MATHEMATICS

COURSE CONTENTS

All of this content will be subject to sampling in the external assessment.

UNIT 1

1 Algebraic Skills

1.1 know and use the notation $n!$, ^nC_r and $\begin{pmatrix} n \\ r \end{pmatrix}$ ☐

1.2 know the results $\begin{pmatrix} n \\ r \end{pmatrix} = \begin{pmatrix} n \\ n-r \end{pmatrix}$ and ☐

$\begin{pmatrix} n \\ r-1 \end{pmatrix} + \begin{pmatrix} n \\ r \end{pmatrix} = \begin{pmatrix} n+1 \\ r \end{pmatrix}$ ☐

1.3 know Pascal's triangle. Pascal's triangle should be extended up to $n = 7$ ☐

1.4 know and use the binomial theorem

$(a+b)^n = \sum_{r=0}^{n} \begin{pmatrix} n \\ r \end{pmatrix} a^{n-r} b^r$, for $r, n \in N$ ☐

e.g. expand $(2u - 3v)^5$ [A/B] ☐

1.5 evaluate specific terms in a binomial expansion,

e.g. x^5 in $(x+3)^7$ **e.g. x^7 in $\left(x + \dfrac{2}{x} \right)^9$ [A/B]** ☐

1.6 express a proper rational function as a sum of partial fractions (denominator of degree at most 3 and easily factorised) ☐

include cases where an improper rational function is reduced to a polynomial and a proper rational function by division or otherwise [A/B] ☐

2 Rules of Differentiation

2.1 know the meanings of the terms 'limit', 'derivative', 'differentiable at a point', 'differentiable on an interval', 'derived function' and 'second derivative' ☐

2.2 use the notation: $f'(x), f''(x), \dfrac{dy}{dx}, \dfrac{d^2y}{dx^2}$ ☐

2.3 recall the derivatives of $x^\alpha (\alpha \text{ rational})$, $\sin x$ and $\cos x$ ☐

2.4 know and use the rules for differentiating linear sums, products, quotients and composition of functions:

$(f(x) + g(x))' = f'(x) + g'(x)$; ☐

$(kf(x))' = kf'(x)$, where k is a constant; ☐

the chain rule: $(f(g(x)))' = f'(g(x))g'(x)$; ☐

the product rule: $(f(x)g(x))' = f'(x)g(x) + f(x)g'(x)$; ☐

the quotient rule:

$\left(\dfrac{f(x)}{g(x)} \right)' = \dfrac{f'(x)g(x) - f(x)g'(x)}{(g(x))^2}$ ☐

differentiate given functions which require more than one application of one or more of the chain rule, product rule and the quotient rule [A/B] ☐

2.5 know

- the derivative of $\tan x$ ☐
- the definitions and derivatives of $\sec x$, $\operatorname{cosec} x$ and $\cot x$ ☐
- the derivatives of $e^x (\exp x)$ and $\ln x$ ☐

2.6 know the definition

$f'(x) = \lim\limits_{h \to 0} \dfrac{f(x+h) - f(x)}{h}$ ☐

2.7 know the definition of higher derivatives

$f^n(x), \dfrac{d^n y}{dx^n}$ ☐

2.8 apply differentiation to:

 (i) rectilinear motion ☐

 (ii) extrema of functions: the maximum and minimum values of a continuous function f defined on a closed interval $[a, b]$ can occur at stationary points, **end points or points where f' is not defined [A/B]** ☐

 (iii) optimisation problems ☐

3 Integration

3.1 know the meanings of the terms 'integrate', 'integrable', 'integral', 'indefinite integral', 'definite integral' and 'constant of integration' ☐

3.2 recall standard integrals of x^α ($\alpha \in Q$, $\alpha \neq -1$), $\sin x$ and $\cos x$ and know the following:

$\int (af(x) + bg(x))dx = a\int f(x)dx + b\int g(x)dx$,

$\quad a, b \in R$ ☐

$\int_a^b f(x)dx = \int_a^c f(x)dx + \int_c^b f(x)dx$, $a < c < b$ ☐

$\int_b^a f(x)dx = -\int_a^b f(x)dx$, $b \neq a$ ☐

$\int_a^b f(x)dx = F(b) - F(a)$, where $F'(x) = f(x)$ ☐

3.3 know the integrals of e^x, x^{-1}, $\sec^2 x$ ☐

3.4 • integrate by substitution: expressions requiring a simple substitution ☐

• candidates are expected to integrate simple functions on sight ☐

• expressions where the substitution will be given, e.g. $\int \cos^3 x \sin x \, dx$, $u = \cos x$ ☐

e.g. $\int \dfrac{6\sin x}{\sqrt{1 - 4\cos^2 x}} \, dx$, $u = \cos x$ [A/B] ☐

the following special cases of substitution:

$\int f(ax+b)\,dx$, $\int \dfrac{f'(x)}{f(x)}dx$ ☐

3.5 use an elementary treatment of the integral as a limit using rectangles ☐

3.6 apply integration to the evaluation of areas **including integration with respect to y. Other applications may include** ☐

(i) volumes of simple solids of revolution (disc/washer method) ☐

(ii) speed/time graph [A/B] ☐

4 Properties of Functions

4.1 know the meanings of the terms 'function', 'domain', 'range', 'inverse function', 'critical point', 'stationary point', 'point of inflexion', 'concavity', 'local maxima and minima', 'global maxima and minima', 'continuous', 'discontinuous' and 'asymptote' ☐

4.2 determine the domain and the range of a function ☐

4.3 use the derivative tests for locating and identifying stationary points i.e. concave up $\Leftrightarrow f''(x) > 0$, concave down $\Leftrightarrow f''(x) < 0$, a necessary and sufficient condition for a point of inflexion is a change in concavity ☐

4.4 sketch the graphs of $\sin x$, $\cos x$, $\tan x$, e^x, $\ln x$ and their inverse functions, simple polynomial functions ☐

4.5 know and use the relationship between the graph of $y = f(x)$ and the graphs of $y = kf(x)$, $y = f(x) + k$, $y = f(x + k)$, $y = f(kx)$, where k is a constant ☐

4.6 know and use the relationship between the graph of $y = f(x)$ and the graphs of $y = |f(x)|$, $y = f^{-1}(x)$ ☐

4.7 given the graph of a function f, sketch the graph of a related function ☐

4.8 determine whether a function is symmetrical, even or odd or neither and use these properties in graph sketching ☐

4.9 sketch graphs of real rational functions using available information, derived from calculus and/or algebraic arguments, on zeros, asymptotes (vertical and non-vertical), critical points, symmetry ☐

5 Systems of Linear Equations

5.1 use of the introduction of matrix ideas to organise a system of linear equations ☐

5.2 know the meanings of the terms 'matrix', 'element', 'row', 'column', 'order of a matrix' and 'augmented matrix' ☐

5.3 use elementary row operations (EROs); reduce to upper triangular form using EROs ☐

5.4 solve a 3×3 system of linear equations using Gaussian elimination on an augmented matrix ☐

5.5 find the solution of a system of linear equations $Ax = b$, where A is a square matrix, include cases of unique solution, **no solution (inconsistency) and an infinite family of solutions [A/B]** ☐

5.6 **know the meaning of the term 'ill-conditioned' [A/B]** ☐

5.7 **compare the solutions of related systems of two equations in two unknowns and recognise ill-conditioning [A/B]** ☐

ADVANCED HIGHER MATHEMATICS

1 Further Differentiation

1.1 know the derivatives of $\sin^{-1}x$, $\cos^{-1}x$, $\tan^{-1}x$ ☐

1.2 differentiate any inverse function using the technique:
$$y = f^{-1}(x) \;\Rightarrow\; f(y)$$
$$= x \;\Rightarrow\; (f^{-1}(x))'f'(y) = 1 \text{ etc.}$$ ☐

know the corresponding result $\dfrac{dy}{dx} = \dfrac{1}{dx/dy}$ **[A/B]** ☐

1.3 understand how an equation $f(x, y) = 0$ defines y implicitly as one (or more) function(s) of x ☐

1.4 use implicit differentiation to find first **and second derivatives [A/B]** ☐

1.5 **use logarithmic differentiation, recognising when it is appropriate in extended products and quotients and indices involving the variable [A/B]** ☐

1.6 understand how a function can be defined parametrically ☐

1.7 understand simple applications of parametrically defined functions, e.g. $x^2 + y^2 = r^2$, $x = r\cos\theta$, $y = r\sin\theta$ ☐

1.8 use parametric differentiation to find first **and second derivatives [A/B]**, and apply to motion in a plane ☐

1.9 apply differentiation to related rates in problems where the functional relationship is given explicitly or implicitly ☐

1.10 **solve practical related rates by first establishing a functional relationship between appropriate variables [A/B]** ☐

2 Further Integration

2.1 know the integrals of $\dfrac{1}{\sqrt{1-x^2}}$, $\dfrac{1}{1+x^2}$; use the ☐

substitution $x = at$ to integrate functions of the form $\dfrac{1}{\sqrt{a^2 - x^2}}$, $\dfrac{1}{a^2 + x^2}$; integrate rational ☐

functions, both proper and **improper [A/B]**, by means of partial fractions, the degree of the denominator being $\leqslant 3$; the denominator may include: ☐

(i) two separate or repeated linear factors ☐

(ii) three linear factors with constant numerator **and with non-constant numerator [A/B]** ☐

(iii) **a linear factor and an irreducible quadratic factor of the form $x^2 + a$ [A/B]** ☐

2.2 integrate by parts with one application ☐

2.3 **integrate by parts involving repeated applications [A/B]** ☐

2.4 know the definition of a differential equation and the meanings of the terms 'linear', 'order', 'general solution', 'arbitrary constants', 'particular solution' and 'initial condition' ☐

2.5 solve first-order differential equations (variables separable) ☐

2.6 formulate a simple statement involving rate of change as a simple separable first-order differential equation, including the finding of a curve in the plane, given the equation of the tangent at (x, y), which passes through a given point ☐

2.7 know the laws of growth and decay: applications in practical contexts ☐

3 Complex Numbers

3.1 know the definition of i as a solution of $x^2 + 1 = 0$, so that $i = \sqrt{-1}$ ☐

3.2 know the definition of the set of complex numbers as $C = \{a + ib : a, b \in R\}$ ☐

3.3 know the definition of real and imaginary parts ☐

3.4 know the terms 'complex plane' and 'Argand diagram' ☐

3.5 plot complex numbers as points in the complex plane ☐

3.6 perform algebraic operations on complex numbers: equality (equating real and imaginary parts), addition, subtraction, multiplication and division ☐

3.7 evaluate the modulus, argument and conjugate of complex numbers ☐

3.8 convert between Cartesian and polar form ☐

3.9 know the fundamental theorem of algebra and the conjugate roots property ☐

3.10 factorise polynomials with real coefficients ☐

3.11 solve simple equations involving a complex variable by equating real and imaginary parts ☐

e.g. solve $z + i = 2\bar{z} + 1$,   **solve $z^2 = 2\bar{z}$ [A/B]** ☐

3.12 interpret geometrically certain equations or inequalities in the complex plane, e.g. $|z| = 1$; $|z - a| = b$; $|z - 1| = |z - i|$; $|z - a| > b$ **[A/B]** ☐

3.13 know and use de Moivre's theorem with positive integer indices **and fractional indices [A/B]** ☐

3.14 **apply de Moivre's theorem to multiple angle trigonometric formulae [A/B]** ☐

3.15 **apply de Moivre's theorem to find nth roots of unity [A/B]** ☐

4 Sequences and Series

4.1 know the meanings of the terms 'infinite sequence', 'infinite series', 'nth term', 'sum to n terms' (partial sum), 'limit', 'sum to infinity' (limit to infinity of the sequence of partial sums), 'common difference', 'arithmetic sequence', 'common ratio', 'geometric sequence' and 'recurrence relation' ☐

4.2 know and use the formulae $u_n = a + (n-1)d$ and $S_n = \frac{1}{2}n[2a+(n-1)d]$ for the nth term and the sum to n terms of an arithmetic series, respectively ☐

4.3 know and use the formulae $u_n = ar^{n-1}$ and $S_n = \frac{a(1-r^n)}{1-r}$, $r \neq 1$, for the nth term and the sum to n terms of a geometric series, respectively ☐

4.4 know and use the condition on r for the sum to infinity to exist and the formula $S_\infty = \frac{a}{1-r}$ for the sum to infinity of a geometric series where $|r| < 1$ **[A/B]** ☐

4.5 expand $\frac{1}{1-r}$ as a geometric series **and tend to $\frac{1}{a+b}$ [A/B]** ☐

4.6 know the sequence $\left(1 + \frac{1}{n}\right)^n$ and its limit ☐

4.7 know and use the Σ notation ☐

4.8 know the formula $\sum_{r=1}^{n} r = \frac{1}{2}n(n+1)$ and apply it ☐ to simple sums, e.g. $\sum_{r=1}^{n}(ar+b) = a\sum_{r=1}^{n}r + \sum_{r=1}^{n}b$ ☐

5 Number Theory and Proof

5.1 understand the nature of mathematical proof ☐

5.2 understand and make use of the notations $\Rightarrow$, $\Leftarrow$ and $\Leftrightarrow$; know the corresponding terminology 'implies', 'implied by', 'equivalence' ☐

5.3 know the terms 'natural number', 'prime number', 'rational number' and 'irrational number' ☐

5.4 know and use the fundamental theorem of arithmetic ☐

5.5 disprove a conjecture by providing a counterexample ☐

5.6 use proof by contradiction in simple examples ☐

5.7 use proof by mathematical induction in simple examples ☐

5.8 prove the following results: $\sum_{r=1}^{n} r = \frac{1}{2}n(n+1)$; ☐

the binomial theorem for positive integers; ☐

de Moivre's theorem for positive integers ☐

ADVANCED HIGHER MATHEMATICS

UNIT 3

1 Vectors in Three Dimensions

1.1 know the meanings of the terms 'position vector', 'unit vector', 'scalar triple product', 'vector product', 'components' and 'direction ratios/cosines'

1.2 calculate scalar and vector products in three dimensions

1.3 know that $\mathbf{a} \times \mathbf{b} = -\mathbf{b} \times \mathbf{a}$

1.4 find $\mathbf{a} \times \mathbf{b}$ and $\mathbf{a}.\mathbf{b} \times \mathbf{c}$ in component form

1.5 know the equation of a line in vector form, parametric and symmetric form

1.6 know the equation of a plane in vector form, parametric and symmetric form, Cartesian form

1.7 find the equations of lines and planes given suitable defining information

1.8 **find the angles between two lines, between two planes and between a line and a plane [A/B]**

1.9 **find the intersection of two lines, a line and a plane and two or three planes [A/B]**

2 Matrix Algebra

2.1 know the meanings of the terms 'matrix', 'element', 'row', 'column', 'order', 'identity matrix', 'inverse', 'determinant', 'singular', 'non-singular' and 'transpose'

2.2 perform matrix operations: addition, subtraction, multiplication by a scalar, multiplication, establish equality of matrices

2.3 know the properties of the operations: $A + B = B + A$; $AB \neq BA$ in general; $(AB)C = A(BC)$;

$A(B + C) = AB + AC$; $(A')' = A$; $(A + B)' = A' + B'$; $(AB)' = B'A'$; $(AB)^{-1} = B^{-1}A^{-1}$; $\det(AB) = \det A \det B$

2.4 calculate the determinant of 2×2 and 3×3 matrices

2.5 know the relationship of the determinant to invertibility

2.6 find the inverse of a 2×2 matrix

2.7 find the inverse, where it exists, of a 3×3 matrix by elementary row operations

2.8 know the role of the inverse matrix in solving linear systems

2.9 use 2×2 matrices to represent geometrical transformations in the (x, y) plane

3 Further Sequences and Series

3.1 know the term 'power series'

3.2 understand and use the Maclaurin series:

$$f(x) = \sum_{r=0}^{\infty} \frac{x^r}{r!} f^{(r)}(0)$$

3.3 find the Maclaurin series of simple functions: e^x, $\sin x$, $\cos x$, $\tan^{-1}x$, $(1 + x)^{\alpha}$, $\ln(1 + x)$, knowing their range of validity

3.4 find the Maclaurin expansions for simple composites, e.g. e^{2x}, **e.g. $e^{\sin x}$, $e^x \cos 3x$ [A/B]**

3.5 use the Maclaurin series expansion to find power series for simple functions to a stated number of terms

3.6 use iterative schemes of the form $x_{n+1} = g(x_n)$, $n = 0, 1, 2, \ldots$ to solve equations where $x = g(x)$ is a rearrangement of the original equation

3.7 use graphical techniques to locate an approximate solution x_0

3.8 know the condition for convergence of the sequence $\{x_n\}$ given by $x_{n+1} = g(x_n)$, $n = 0, 1, 2, \ldots$

4 Further Differential Equations

4.1 solve first-order linear differential equations using the integrating factor method

4.2 find general solutions and solve initial value problems

4.3 know the meanings of the terms 'second-order linear differential equation with constant coefficients', 'homogeneous', 'non-homogeneous', 'auxiliary equation', 'complementary function' and 'particular integral'

4.4 solve second-order homogeneous ordinary differential equations with constant coefficients

$$a\frac{d^2 y}{dx^2} + b\frac{dy}{dx} + cy = 0$$

4.5 find the general solution in the three cases where the roots of the auxiliary equation:

(i) are real and distinct

(ii) **coincide (are equal) [A/B]**

(iii) **are complex conjugates [A/B]**

4.6 solve initial-value problems

4.7 **solve second-order non-homogeneous ordinary differential equations with constant coefficients**

$$a\frac{d^2 y}{dx^2} + b\frac{dy}{dx} + cy = f(x) \text{ using the auxiliary}$$

equation and particular integral method [A/B]

5 Further Number Theory and Proof

5.1 know the terms 'necessary condition', 'sufficient condition', 'if and only if', 'converse' and 'negation'

5.2 use further methods of mathematical proof: some simple examples involving the natural numbers

5.3 direct methods of proof: sums of certain series and other straightforward results

5.4 further proof by contradiction

5.5 further proof by mathematical induction prove the following result

$$\sum_{r=1}^{n} r^2 = \frac{1}{6}n(n+1)(2n+1); \; n \in N$$

5.6 know the result $\sum_{r=1}^{n} r^3 = \frac{1}{4}n^2(n+1)^2$

5.7 apply the above results and the one for $\sum_{r=1}^{n} r$ to prove by direct methods results concerning other sums

e.g. $\sum_{r=1}^{n} r(r+1) = \frac{1}{3}n(n+1)(n+2)$

e.g. $\sum_{i=1}^{4} (3i+1)$

e.g. $\sum_{r=1}^{n} r(r+1)(r+2) = \frac{1}{4}n(n+1)(n+2)(n+3)$

e.g. $\sum_{r=1}^{n} r(r^2+2)$ [A/B]

5.8 know the division algorithm and proof

5.9 use Euclid's algorithm to find the greatest common divisor (g.c.d.) of two positive integers

5.10 **know how to express the g.c.d. as a linear combination of the two integers [A/B]**

5.11 **use the division algorithm to write integers in terms of bases other than 10 [A/B]**